Mark Anthony Benvenuto
Chemistry and Energy

W0235434

Also of Interest

Energy and Sustainable Development
Warren, 2021
ISBN 978-1-5015-1973-4, e-ISBN 978-1-5015-1977-2

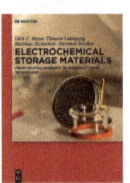

Electrochemical Storage Materials.
From Crystallography to Manufacturing Technology
Meyer, Leisegang, Zschornak, Stöcker (Eds.), 2019
ISBN 978-3-11-049137-1, e-ISBN 978-3-11-049398-6

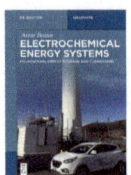

Electrochemical Energy Systems.
Foundations, Energy Storage and Conversion
Braun, 2019
ISBN 978-3-11-056182-1, e-ISBN 978-3-11-056183-8

X-ray Studies on Electrochemical Systems.
Synchrotron Methods for Energy Materials
Braun, 2017
ISBN 978-3-11-043750-8, e-ISBN 978-3-11-042788-2

Organic and Hybrid Solar Cells.
An Introduction
Schmidt-Mende, Weickert, 2016
ISBN 978-3-11-028318-1, e-ISBN 978-3-11-028320-4

Mark Anthony Benvenuto

Chemistry and Energy

From Conventional to Renewable

DE GRUYTER

Author
Prof. Dr. Mark Anthony Benvenuto
Department of Chemistry and Biochemistry
University of Detroit Mercy
4001 W. McNichols Rd.
Detroit, MI 48221-3038
United States of America
benvenma@udmercy.edu

ISBN 978-3-11-066226-9
e-ISBN (PDF) 978-3-11-066227-6
e-ISBN (EPUB) 978-3-11-066233-7

Library of Congress Control Number: 2021944975

Bibliographic information published by the Deutsche Nationalbibliothek
The Deutsche Nationalbibliothek lists this publication in the Deutsche Nationalbibliografie;
detailed bibliographic data are available on the Internet at http://dnb.dnb.de.

© 2022 Walter de Gruyter GmbH, Berlin/Boston
Cover image: 1715d1db_3/iStock/Getty Images Plus
Typesetting: Integra Software Services Pvt. Ltd
Printing and binding: CPI books GmbH, Leck

www.degruyter.com

Contents

Chapter 4
Biofuels —— 43

Chapter 5
Nuclear power —— 53

Chapter 9
Wind power —— 95

Chapter 10
Energy storage —— 106

Chapter 11
Energy harvesting —— 119

Chapter 1
Introduction

1.1 History

The history of humanity is tied up with the generation and use of energy. The world we live in today still uses some types of energy that it has used since antiquity, such as fire to cook food, even, if only at a summer outdoor gathering. But, obviously, there are numerous energy sources in use today that have only been harnessed and come into use in the past two centuries, and some that are much newer than that.

1.1.1 The earliest conversion of materials to energy

Even before humans had stopped being hunter-gatherers, we had started using and depending on a steady supply of energy beyond that furnished by our own muscles. Taming wood fires, the beginnings of which are lost in some ancient past, allowed people to cook food; and thus a wide variety of woods, grasses, and animal dungs were used in that process, and concurrently in keeping humans warm. In addition, as humans began to herd animals, the warmth of those animals was used to keep people and their dwellings warm in the evenings and in the cold seasons of the year. Importantly, the muscles of animals proved to be useful in doing work that would have been done by humans earlier – work such as pulling a plow, a wagon, or a sled.

By the time humanity had begun to settle into permanent living areas – those communities which would eventually become cities – fire had been harnessed not only to keep people and animals warm but to smelt metals and to fire clay. The invention of both the kiln and the forge, and the use of coal to fire them, allowed advances in human life and the quality of that life precisely because the energy of fire was channeled into the production of specific materials, such as pottery and metal tools or weapons. Some modern artisans and craftsmen still utilize these ancient techniques, an example of which is seen in Figure 1.1. In turn, fire was generated by the combustion of woods and animal dungs – and at times by the hotter combustion of coal. It is not an exaggeration to claim that the use of fire in this manner helped bring certain peoples to the fore, and ultimately helped bring about the rise of city-states and empires.

The use of fire and its traditional fuels on a large scale had consequences that could not have been seen at the time it began. For example, the use of wood fires to heat homes at the start of the Industrial Revolution meant that some cities had an almost constant haze about them, and smelled as if there was some never-ending bonfire ablaze. Such use also deforested areas, sometimes large ones. In England, for example, for many years, it was against the law to fell large trees for such things as

https://doi.org/10.1515/9783110662276-001

Figure 1.1: Bronze sword.

firewood, as these were reserved for use in ship building by the Royal Navy. Farther back, the use of coal to smelt metals rose to a large enough extent during the Roman Empire that it polluted a large swath of Europe, as far away as Greenland. A fascinating study was undertaken in the 1980s, in which ice cores of the Greenland ice sheets were examined to determine pollutant inclusions over a broad period of time. It was found that the by-products of incomplete coal combustion and metal refining were located in strata of the cores that were laid down during the time Rome rose to prominence [1].

But wood, animal dung, and coal were not the only sources of energy used in the distant past. Sailing ships and windmills certainly used the power of the circulating air, and the sun's rays were harnessed to dry food, produce salt from brine ponds, and aid humans in several other ways. In addition, dams have a long history, both for storing water and for controlling its use [2]. The power of moving water has been used to turn wheels and thus do some needed task for over millennia [3].

1.1.2 Solar

The idea of concentrated solar power is a relatively new one, in terms of producing energy, but also has an interesting historical footnote. According to legend, the famous Archimedes in the third century BC directed the use of polished bronze shields against a fleet of invading Roman ships. Supposedly the shields were directed so that the sun's rays were concentrated at the ships, which eventually caught fire. The story, while a colorful and impressive example of a possibility for the use of concentrated solar power, is doubted by many historians today.

One chapter in solar energy that is not in doubt is the much more modern first solar collector patent, issued to Alessandro Battaglia, in 1886 [4]. Since then, solar power has continued to find applications, either for generating small amounts of electricity, or large. But this also continues to struggle economically against the prices of coal and oil when it comes to feedstocks for materials that can generate the needed power.

1.1.3 Water wheels

Curiously, the idea of using a water wheel to produce and harness the power of flowing water goes far back before the use of it to generate electricity. Archaeologists and historians believe that the earliest water wheel, one driven by tidal water power, was at the Nendrum Monastery mill in Northern Ireland, and was in use as early as the year 787 CE. This remains a subject of debate, as water wheels have been mentioned in the writings of the ancient Greeks and Romans. Perhaps obviously, the use of any configuration of water wheel to generate electricity is a much more recent accomplishment [3].

1.1.4 Wind

Wind power has a very old history as well, but for most of it wind was only used for some single mill or for a sailing ship. Much more recently, wind has been used to generate power for a larger, integrated power grid. Figure 1.2 shows part of a modern wind farm in southwest Ontario. Each tower rises several stories higher than the average home, and all are linked to the local power grid. The construction of such windmills and their internal mechanisms requires materials that have never before been required in such large quantities. An example is lanthanides, such as neodymium, which are used in motors of such windmills.

Figure 1.2: Wind farm.

1.2 The industrial revolution

There have been only a few points in history that one specific event or change in thinking has affected the world in large and all-encompassing ways. One can argue that the invention and spread of laws is one such event – a means by which all people in a community are expected to act and behave in the same manner, and for which standard punishments are meted out for transgressions. The initial use and continuous spread of coins may be another – the idea that some object or thing can be used as a common means of value for a variety of other objects of value, even when those objects are not located in the same place. But by far, the inception of what we now call the Industrial Revolution is one of those events that have changed history, on par with the long-term storage of different foods, taming of the horse, and even the rise of the Roman Empire.

The Industrial Revolution appears to have its origin in Great Britain. There remains debate as to what the dominant factor or factors were that started it, whether it was a surge in human population, or changes in farming, or changes in climate that brought about these just-mentioned two, some combination of the above three, or some other factors that we have not mentioned. What is not in doubt is that England sits geologically on one enormous bed of coal, one that could be reached with mining techniques that were known three centuries ago [4]. This buried material did become a source of power when it was burned in order to heat water, and the resulting steam is used to obtain useful work from steam turbines. One other source of power at the time was the water wheel.

Textiles were one of the markets that rapidly expanded during the Industrial Revolution, and the new mechanical looms and other machinery that produced larger amounts of thread, cloth, and finished products all required energy. And while it may seem odd that this was all needed even though the size of the British population, and thus the consumers for such goods, had not increased dramatically, it is critical to remember that Britain was the center of an enormous, world-spanning empire. Products made there were sold to people throughout the British Empire, as well as to businesses and people in other countries. Bigger markets were therefore accommodated, and thus more energy was required.

Certainly, other commodities were involved in the expanding of production during the Industrial Revolution, including several commodity chemicals. But one of those that has been tracked and studied extensively is the production of iron.

Iron had been produced in relatively small quantities since ancient times, but the Industrial Revolution both expanded this production and at the same time expanded the uses for it. Important developments were made in forges and furnaces, including a wider spread use of Bessemer furnaces, and in the use of coke as a reducing agent as opposed to charcoal. The adoption of coke for the production of reduced iron is a chemical development that remains with us and the iron and steel industry today.

Along with these developments came improvements in transportation related to moving large volumes of products and materials to markets. These too require energy, such as that needed to drive locomotives, and that required to dig canals – and then to harness the power of moving and falling water. Although large-scale transport of goods via canal has certainly declined to essentially nothing, the continued use of railroads remains an area for which energy is required, and for which it is linked to other industries.

1.2.1 The human potential and cost

There have undoubtedly been costs associated with the inception and expansion of the Industrial Revolution, some of which have been horrible – such as working conditions for employees in some industries – and some of which remain with us today – such as regulating daily work time with clocks. Machines were and are in a way served by people, and so the idea of shift work, connected to specific time on increasingly specific clocks, came into existence. This allowed machines to run constantly, which meant that energy was needed constantly, and which also meant that workers sometimes worked in toxic, unsafe conditions to keep the sources of energy available and to keep the machinery running. The idea of days being regulated by the sun and by church bells – which worked for centuries in largely agrarian communities and societies – was not conducive to the regularity of machines and the energy input needed to keep them functioning. Every time we use an alarm clock today, we participate in the modern version of this machine-and-energy-driven behavior. But also, as more and more industrialized processes are initiated, people realize that the use of ever-more energy is simply not possible. This, and the cost of the materials that produce energy, continued to drive efforts at energy efficiency.

1.3 Industrialized nations

The industrialized nations of the world today can be defined in a number of ways, one might be as those with large and continued productivity of several tens of thousands of chemicals, many of them derived from petroleum. Until the end of the Second World War, the energy required for this scale of production was provided almost exclusively by the burning of coal or oil, with a certain amount of hydroelectric power as well, in the form of dammed rivers. In the past 30 years, the profile of energy sources has widened considerably, to include more and different kinds of hydroelectric power but also to include the power of controlled nuclear fission, as well as that derived from wind and the Sun. These newer forms of power have all been developed in the past 70 years, and some remain controversial to the present. Figure 1.3 shows an aerial view of a nuclear power plant, its cooling towers, and their distinctive shape being prominent. Despite its decades of use, some individuals remain deeply skeptical of nuclear power.

Figure 1.3: Aerial view of a nuclear power plant.

1.4 Energy and quality of life

Discussions of what gets called quality of life often revolve around how long a person or group of people live, or what level of medical and health care they have available to them throughout their lives. But a deeper examination of the idea routinely comes back to how readily available energy is in some easy-to-access form for a person or group of people. People who live without electricity or automotive transport often do not have the ability to store food for long periods of time, or to get to food that is not locally grown, or to get access to certain medicines and advanced health care. As well, health-care facilities routinely depend on electricity for virtually all their tests and procedures. On the other hand, enormous numbers of applications of energy, in the form of electricity or otherwise, are often taken for granted by those who have constant access to them. For example, Figure 1.4 shows a frozen food aisle in a grocery store. Electricity is used continuously in this application, every day, without stop, simply to keep the unseen heat exchangers running that keeps a multitude of foods cold. Without all of these applications of energy, we would not live at the quality of life that has become common today.

Figure 1.4: Frozen food aisle.

1.4.1 Automotive transport

The idea of motorized, automotive transport is now common to every person in the developed world, but this remains a relatively new development in history. With this comes a very different use of materials for energy, one that often dominates local or area economies.

Prior to the use of automobiles, there was certainly mechanized transport, in the form of trains on land, and steam ships by water. But prior to these developments, the fastest a person could travel on land was by horse, and the fastest by sea was via a sailing ship. This had not changed from the time of the earliest of the ancient civilizations such as those in Mesopotamia, roughly 3,500 BC, to the time of Napoleon, at the beginning of the nineteenth century. From that latter time until the time of the United States Civil War, 1861–1865, humanity advanced from horse and sail to rail and steam ship, both of which saw continuous improvements over the course of time, an example of which is shown in Figure 1.5. But these advancements required energy, almost exclusively in the form of coal.

From the US Civil War to the First World War, only 49 years, humanity advanced to personal automotive transport and mechanized airplanes. And from the point at the end of the First World War, the year 1918, until 1969, when a person first stepped on the Moon, humanity advanced to enormous ore-hauling ships, as shown in Figure 1.6, jet-powered flight, and the use of rockets. All these advances

Figure 1.5: Modern passenger train.

Figure 1.6: Ore hauling, ocean-going transport ship.

required the continued use and harnessing of chemical energy, often in enormous, never-before needed amounts.

1.4.2 Mass communication

Much like transportation, the development of mass communication is tightly intertwined with the availability of and the use of energy. Throughout almost the entire span of human history, for thousands of years, the written letter has been the main form of long-distance communication between two individuals who were far enough away that they could not see each other. Eventually, printed newspapers and mechanically printed books became a major means of communication. The machinery that enabled these two forms of writing did require more energy than any handwritten document, although such energy output was small compared to the output that could be achieved, and compared to what would come next.

In the mid-1700s, the first telegraph machinery was produced and proven to be useful to send messages. However, what can be called the first electrical telegraphs date only to the early nineteenth century. This represented significant progress in terms of communication, but once again required more energy.

The invention of the telephone further revolutionized communication – Bell was granted his first patent for such a device in 1876 – and further increased the amount of energy required to enable this new form of communication [6]. While different forms of what were then called telephones did exist prior to Bell's inventions, they were not always electrical. The adoption and spread of electrical telephones created a network of devices that fed from a growing set of power grids, grids that are now national and international in scope.

Perhaps obviously, the greatest advance in communication has only come in the last few decades – the internet. It is now possible to download huge amounts of data, full movies, entire books, and a host of other files, all because they are available from some server, and all because a power grid that is large and extensive enough that it can power all the computers required exists.

1.4.3 The internet

The modern internet appears to have its origins in the military alliance that is NATO, in the late 1960s and early 1970s. Efficient, effective communication between the military leaders of the member nations, or more specifically, between their computers, was considered important as part of the defense posture needed to prevent a war with the then Soviet Union. When this first set of connections was created, it too required some energy input – and a dedicated source of electricity was considered essential to the security of the growing network. But this idea of connected computers obviously did not stay exclusive to military organizations.

Interestingly, as computers evolved and their abilities expanded, the energy need or requirement for any single computer decreased. But the enormous growth of the number of computers, and later of cell phones and other electronic devices that incorporated significant computing power, has meant that the energy requirements for the whole has continued to increase. Perhaps obviously, the internet that exists today has far outstripped the original vision of any such network by orders of magnitude. All of this requires a steady, constant flow of energy.

The growth of the internet has led to a certain amount of speculation about the energy required to use and maintain it. The general numbers now are frankly amazing [5]. More than one billion smartphones, one billion laptop computers, and almost one billion desktop computers have been manufactured, and while all of them are not still in service, all that are do require electricity to function. This is taken from existing electrical grids, and thus adds a burden to them beyond what might be called normal residential, commercial, and industrial use. As well, the number of servers continues to grow – currently estimated at approximately 80 million. Again, these require electricity on a constant basis.

Overall estimates from a variety of sources are that the internet uses between 3½% and 6% of the electricity produced globally, not wildly different from early estimates [5]. While we have just mentioned everyday devices, including personal computers and the data centers where servers reside, there are also power requirements in terms of buildings in which they are located, which also require constant energy to keep them temperature controlled, meaning cool enough that they can continue to function without interruption, 24 h a day, every day of the year.

Clearly, these requirements have become enormous, have secondary and tertiary equipment and infrastructure connected to them which also requires electricity, and have continued to grow since this first studied estimate of energy use by the internet was formulated.

1.4.4 Creature comforts

There is no doubt or question that the average person today has at his or her grasp a much wider array of objects that can be called labor saving devices – often objects that require some form of electrical power – than even kings, queens, or the wealthiest of society did a century or more ago. Electric toothbrushes, razors, hairdryers, garage door openers, electric stoves, forced air home heating, all these and many, many more normal devices of life today were unknown a century ago; and those listed represent only a tiny sampling of what exists today. Almost all of these require energy in some form.

The transport of a wide variety of liquid fuels is another aspect of modern life that enables the quality of life we live today, an example of which is shown in Figure 1.7. Pipelines exist that cross huge expanses of land, which split into smaller pipelines, which in turn split into lines where petroleum products can be stored in tanks. From here, such products can be shipped by rail car or by truck to commercial gas stations, where they can be purchased by consumers. This level of construction and organization, all for the distribution of energy, is unlike anything that has been seen throughout history.

In addition, the ability to ship food by air, rail, ship, and truck, in refrigerated cars and containers, enables a healthier diet for much of the world than what was available before the advent of inexpensive temperature control. This has increased the life span of people today, although it has not eliminated all dietary health-related problems, and indeed, appears to have caused some. Also, the transport of food always requires some form of electrical or chemical energy [7,8].

Figure 1.7: Truck transport of hydrocarbons.

References

[1] S. Hong, J.-P. Candelone, C.C. Patterson, C.F. Boutron. Greenland ice evidence of hemispheric lead pollution two millennia ago by Greek and Roman civilizations. Science. 265(5180), 1841–1843, 1994.

[2] H. Yang, M. Haynes, S. Winzenread, K. Okada. The History of Dams. Website. (Accessed 31 March 2021, as: https://watershed.ucdavis.edu/shed/lund/dams/Dam_History_Page/History.htm).

[3] P.-L. Viollet. From the water wheel to turbines and hydroelectricity. Technological evolution and revolutions. Comptes Rendus Mécanique. 345(8), August 2017, 570–580.

[4] A. Battaglia. "Sul modo e sulla convenienza di utilizzare il calore solare per le machine a vapore." Paper, by Prof. E. Semmola at the session, April 17th, 1884, Proceedings of the Naples Institute of Encouragement, 1884.

[5] S. Winchester. The Map That Changed the World: William Smith and the Birth of Modern Geology, Harper Collins Publishers, 2001, ISBN: 0-06-019361-1, New York.

[6] Alexander Graham Bell – patent # US174,465A. Improvements in telegraphy.

[7] F. Haggstrom, J. Delsing. IoT energy storage – a forecast. Energy Harvesting and Systems. 2018, 5(3–4), 43–51.

[8] L.M. Hilty, B. Aebischer eds ICT Innovations for Sustainability, chapter, The Energy Intensity of the Internet: Home and Access Networks, 2014, Springer Int'l Pub, 3–36.

Chapter 2
Coal

2.1 Introduction, history, geographic locations

The first use of coal as a fuel has been lost to time, but was confined to that coal found at the surface of the Earth, in a variety of locations worldwide, which can be called placer coal. Historically, enough was gathered or mined from relatively shallow deposits that it was usable for fashioning bronze items and later iron tools and weapons throughout many of the cultures of the world.

Unlike many mined substances, coal is a sedimentary type of rock that is not mineral in composition. It is an organic material, not a mineral (like most types of what can be called rock). Depending on its grade, it is primarily made of carbon, and very importantly, it burns.

The rise of the modern coal mining industry has its origins in the Industrial Revolution, and in England. Geologically, all of Great Britain rests atop an enormous bed of coal, which means that mines can be dug in numerous places. In the popular press, Simon Winchester's book, *The Map That Changed the World* is a thorough treatment of how geologic maps are made, as well as how the presence and placement of coal was determined throughout Great Britain [1]. Along with an ability to chart where coal could be found, such maps changed the human perception of how old the world was, and whether parts of the Old Testament, such as genealogy lists, could be considered literal in terms of the timelines they presented.

While the Industrial Revolution marks the beginning of the large-scale use of coal for the generation of power, coal mining has continued to expand and develop both in size and in efficiency, and is now one of the largest classes of mining operations in the world largely because the amount of coal that is available and usable for power generation is by far the largest potential energy source available. Numerous regional, national, and international organizations exist to promote the use of coal and any related materials [2–34]. Interestingly, this does not necessarily mean that coal mining occupies the most people. Increased efficiency of operations means that this industry currently occupies approximately 50,000 people in the United States, for example. If the series of improvements in equipment and machinery had never taken place, and the same techniques were used today as had been used in the mid-1800s, a much larger work force would be required to produce the amount of coal currently used each year. Still, the mining of coal is an important enough industry that the trade organizations devoted to its promotion and use advocate for the workers and other employees, as well as for the material [2–29].

https://doi.org/10.1515/9783110662276-002

2.2 Grades and combustion of coal

Coal exists always as a mixture of hydrocarbons, oxygenated hydrocarbons, and other compounds, which means it has no fixed chemical composition. This is because its origin is from living matter that has been compressed over millions of years, and thus that started with complex molecules. While different sources show different versions of what they claim is coal's structure, with numerous linked aliphatic and aromatic cycles, such structures change with each sample of coal, especially from different regions throughout the world. Thus, it exists in various grades, and while those are routinely cited in terms of hardness, they can also be correlated to how well specific grades or types combust, and what sort of gaseous effluent by-products they produce. As well, one can consider different grades of coal in terms of the amount of carbon in a particular sample. Table 2.1 shows the broad classes of coal.

Table 2.1: Grades of coal.

Name	Common names	Description	Comment
Anthracite	Hard coal	High amount of fixed carbon	Domestic fuel. Use in automatic stoker furnaces
Bituminous	Black coal	High heat value	Often used to generate electricity, contains methane (aka "firedamp")
Sub-bituminous		Black in color, relatively high heat value	
Lignite	Brown coal	Lowest carbon concentration	Dirtiest emissions

The combustion of coal to generate energy is now a mature industry in many countries. Coal must be fed into a furnace and combusted, while at the same time, water is piped through the furnace in a separate loop to produce steam. The steam is then used to turn a turbine, which is connected to a generator. The generator is then connected to a power grid, and electricity is delivered to the power grid. The steam is often re-cooled to water, and re-routed in a cycle, to be used again. Figure 2.1 shows a basic chemical reaction for the combustion of coal. Figure 2.2 then shows the basic steps of a coal-fired plant operation, while Figure 2.3 shows how a turbine–generator operates. The size of these operations and the need to reintroduce some water into the system at various times are two reasons that coal-fired power plants are often built near some water source.

$$C_xH_y + O_{2(g)} \to^{\Delta} \to xCO_{2(g)} + H_2O_{(g)} + (other)_g$$

Figure 2.1: Combustion of coal.

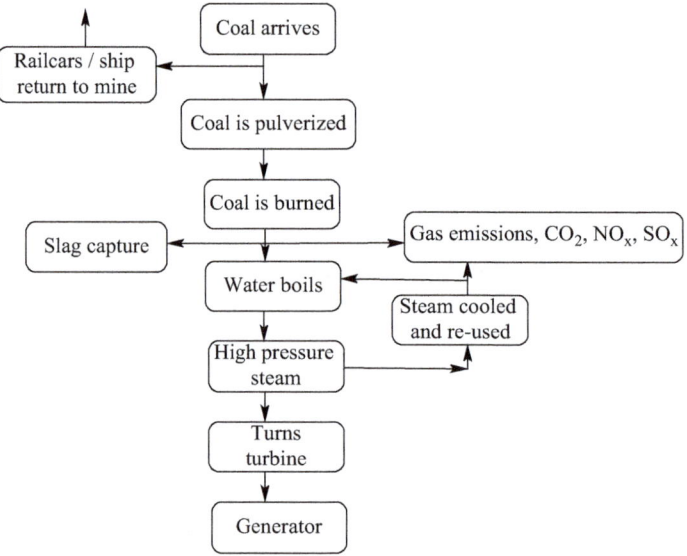

Figure 2.2: Steps from coal to electricity.

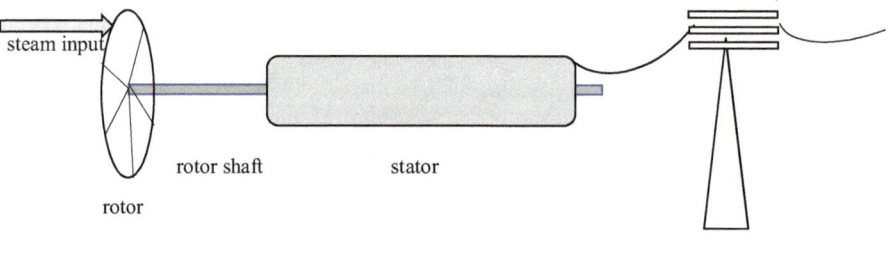

Figure 2.3: Turbine–generator for electricity production.

The basic means by which a turbine and generator produce electricity is actually not very different from the time in the early nineteenth century during which Michael Faraday produced the first generators. Perhaps, obviously, the scale has since become enormous. But the system still depends on a rotor and rotor shaft turning in what is called a stator, the stationary component of any system which has a rotating part. In coal-fired power plants, it is the steam that is generated which makes the rotor turn, and thus the rotor shaft function.

It is noteworthy at this point to mention that in several of the later chapters of the book, what can be called the final step in the production of electricity remains this turbine–generator combination.

The grades of coal that were just discussed thus become important during the production of electricity. Since coal is largely turned into gases in the process, the "other" shown in Figure 2.1 and the chimney of any plant become the main outlet for such gases. When coal that is of higher grade is used, there are less by-products that include sulfur oxides, nitrogen oxides, and, in some cases, mercury. In cases where coal that is high in sulfur or mercury is used, these materials must be separated from gases such as carbon dioxide before release to the atmosphere. Thus, what are called scrubbers are used to minimize the emission of such materials to the atmosphere. Scrubbers are often some absorbent materials such as lime or limestone, with which sulfur oxides, nitrogen oxides, or mercury can combine, and thus not enter the atmosphere.

Additionally, the combustion of coal does produce some solid by-products called coal ash or fly ash, which is shown schematically in Figure 2.2. Since solids are easier to handle and use than are gases, an industry has developed for possible uses for coal ash, so that it is not simply a waste product [30–34]. Coal ash has found its use as pozzolan in the formulation of hydraulic cement, and as a component of other cements. It is also used in the construction of roads, in the sub-base layers. All of these uses can be considered ways of finding use for something that would otherwise simply be landfilled. But it is an impressive feat that so many different applications can be found for what would otherwise be a waste material.

2.3 "Clean" coal

The term "clean coal" has entered the general lexicon in the past few decades and remains hotly debated today. The term encompasses technologies aimed at the removal of several substances prior to combustion, including sulfur oxides, nitrogen oxides, and mercury, or after. Various scrubbers have been developed and utilized to remove these trace materials.

Very importantly, the term clean coal also encompasses the sequestration and capture of carbon dioxide (called carbon capture and sequestration) emitted during coal combustion. This ultimate effluent gas from any coal combustion – carbon dioxide – is not currently eliminated during the combustion of clean coal; hence the continued debate about the term, and the meaning of it. Carbon dioxide is one of the major gases that contribute to global warming and continued climate change. To the credit of researchers examining this phenomenon, there have been chemistries developed to produce oxalates from carbon dioxide, since an oxalate anion is simply a dimer of CO_2, as shown in Figure 2.4. But none of these has yet reached the stage where the process for this production can be scaled up to an industrial level, and done so in an economically feasible manner [35–38].

$$CO_{2(g)} \rightarrow C_2O_4^{2-}{}_{(s)}$$

Figure 2.4: Oxalate formation.

As mentioned, this continued emission of carbon dioxide from any coal-fired power plant remains the main crux of the debate about whether or not there can truly be "clean" coal. Critics argue that without some means of capturing this effluent gas completely, the idea remains only a dream or a goal to be achieved in the future. Others argue that the capture of all effluents other than carbon dioxide will go a long way to what they term "clean coal." Although we have discussed such materials, a listing of what is emitted when coal is burned is shown in Table 2.2.

Table 2.2: Emissions from coal-fired plants.

Emitted	Fate	Comments
Carbon dioxide	To atmosphere	Sequestration efforts are underway in some areas
Fly ash	Captured before release	Sometimes used in road construction
Mercury	Captured/scrubbed	Disposal in designated landfills
Nitrogen oxides (NO_x)	Can be scrubbed	
Particulates	Can be filtered from exhaust	
Sulfur oxides (SO_x)	Can be scrubbed	

2.4 Coal fires

While a great deal of effort and research has gone into the best possible utilization of coal, for energy production as well as for use in steel manufacturing and other industrial operations, the worst "use" of coal is in what are called coal fires. Unfortunately, in several instances, a seam of coal has ignited while still in the earth, and often has proved impossible to extinguish.

Reports indicate that at times thousands of coal fires are burning in different parts of the world. Perhaps the most famous, or infamous, one in the United States is that under Centralia, Pennsylvania. It has been burning since May 27, 1962, appears to be unquenchable, and is estimated to have over two centuries of fuel remaining. The fire appears to have started because people were using one older part of the mine for dumping trash, and trash that had been set on fire was dumped.

More than one attempt has been made to extinguish the fire, by entrenching certain parts of the area, or moving soil and rock, but none of the attempts have proved successful. As a result, the area above the mine has been cleared of people, and the town of Centralia, Pennsylvania, has essentially been abandoned.

There are several coal fires in China as well, which are monitored constantly, to ensure that populations near them are not affected. Additionally, there are coal fires in Australia, Indonesia, India, and different nations in Europe.

Clearly, all coal fires represent a waste of resources and energy, and their combined output in terms of carbon dioxide can be factored into the amount of CO_2 put into the atmosphere by anthropogenic means. Thus far, there has been no standard technique developed for extinguishing a coal fire, simply because each coal deposit is different, and thus each mine layout is in some way different from any other.

2.5 The cycle of carbon

Numerous chemistry or biology textbooks discuss a nitrogen cycle, in terms of fertilizer and uptake by plants, and its interaction as a diatomic element with living organisms. As well, they often discuss an oxygen cycle, usually in terms of respiration and the use of oxygen by living organisms. We can also speak of a carbon cycle, though. This certainly includes coal, but encompasses carbon in the lithosphere, the biosphere, and the atmosphere.

Several different versions of a carbon cycle can be drawn, inclusive or exclusive of coal. The world's oceans are a very large source of carbon to the atmosphere, for example, although their exchange of CO_2 is quite slow. Volcanoes represent another natural means by which carbon is introduced to the atmosphere, in the form of carbon dioxide as well as soot and carbon-containing ash. The three major forests or jungles of the world – the Amazon, the Congo, and the northern boreal forest – are sometimes called the lungs of our planet, and are both a source of CO_2 as well as a sink that consumes it. These natural sources, as well as the direct cycle between all forests and other plant life and atmospheric CO_2, exhibit a rhythm that continues to take place on the planet, and that has done so for possibly billions of years. The exchange of carbon in terms of plant life can be represented broadly by two rather simple reactions, as shown in Figure 2.5.

$6CO_{2(g)} + 6H_2O_{(g\,or\,l)} \rightarrow C_6H_{12}O_{6(s)} + 6O_{2(g)}$

And

$C_6H_{12}O_{6(s)} + 6O_{2(g)} \rightarrow 2C_{(s)} + 2CO_{(g)} + 2CO_{2(g)} + 6H_2O_{(g\,or\,l)}$

Figure 2.5: Carbon exchange with plant life.

The first reaction is the simplified representation of photosynthesis. The second represents the combustion seen in forest fires, or any other event that returns carbon to the atmosphere in a gaseous state or as particulate matter. Note that the carbon-containing products of the second reaction are not simply CO_2. Both carbon monoxide and elemental carbon – the latter in the form of soot – are routinely produced.

However, we represent a carbon cycle, though it is certainly worth noting that the time frame for the production of coal in the Earth and the time frame for the use of it by humans for power generation are vastly different. The just-seen figure is one of the fastest forms of carbon uptake and release. Yet, at a minimum, tens of millions of years are required for the production of coal from living matter – and its combustion and conversion back to CO_2 can happen in a matter of days. Virtually, all the coal that has been consumed on an industrial level, for energy production as well as iron and steel production, has occurred only in the past few hundred years. Figure 2.6 presents a scheme showing the carbon cycle.

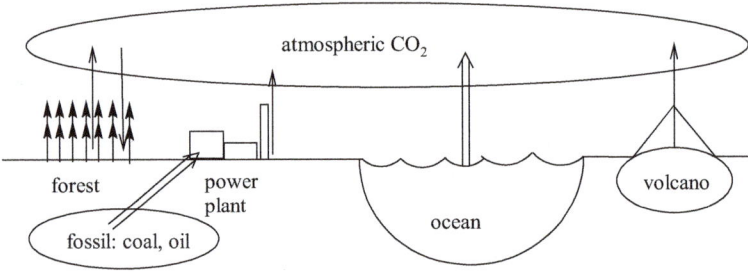

Figure 2.6: Schematic of the carbon cycle.

2.6 Coal emissions and climate change

The emission of so much carbon dioxide into the atmosphere since the beginning of the use of coal for coal-fired power plants has changed our world, even though intense debate still continues among various factions of society and its political leaders about whether such emissions are causing a climate change and global warming or not. The warming occurs because the CO_2 molecule is linear, but has a central atom around which vibration or bending can occur, as shown in Figure 2.7. Molecules such as nitrogen and oxygen, the two main components in the air, cannot vibrate in this manner, and thus cannot absorb energy through some type of bending vibration.

$$O{=}C{=}O \xrightarrow{h\nu} \quad O{=}C_{\diagdown O} \xrightarrow{h\nu} \quad O{=}C^{\diagup O}$$

Figure 2.7: Carbon dioxide vibrations.

In general, the scientific community appears united in the belief that the world is warming, and that temperature extremes are increasing, with the debate being about whether or not we have passed a tipping point, a point of no return, as it were. But numerous persons in positions of political power are not in agreement with this assessment. Their reason is generally not based on the chemistry involved or data collected, but on the political reality of their voting population. It is impossible to tell a community to stop earning money and making living by mining coal, and to stop for several months, while retraining for some other profession. Thus, some political leaders take the position that is best for their constituency, and that the rising levels of CO_2 in the atmosphere is either not caused by the use of coal or is not a problem.

This being said, carbon dioxide in the atmosphere has been tracked extensively at least for the past 60 years. Records indicate that the CO_2 concentration in 1960 was approximately 310 ppm with seasonal variations, and is 420 ppm today [39].

References

[1] S. Winchester. The Map That Changed the World, ISBN: 978-0060193614.
[2] American Coal Council. Website. (Accessed 13 April 2021, as: http://www.americancoalcoun cil.org).
[3] American Coalition for Clean Coal Electricity. Website. (Accessed 13 April 2021, as: http://www.americaspower.org).
[4] Coal Association of Canada. Website. (Accessed 14 March 2021, as: https://coal.ca).
[5] World Coal Organization: Website. (Accessed 13 April 2021, as: https://www.worldcoal.org/).
[6] Coal Association of New Zealand – http://www.straterra.co.nz/industry/coal-association-of-nz/
[7] Australian Coal Industry's Research Program (ACARP). Website. (Accessed 13 April 2021, as: acarp.com.au).
[8] United Mine Workers of America. Website. (Accessed 13 April 2021, as: http://umwa.org).
[9] National Mining Association. Website. (Accessed 13 April 2021, as: https://nma.org).
[10] West Virginia Coal Association. Website. (Accessed 13 April 2021, as: https://www.wvcoal.com).
[11] Friends of Coal. Website. (Accessed 13 April 2021, as: https://www.friendsofcoal.org).
[12] American Coal Foundation. Website. (Accessed 13 April 2021, as: http://www.teachcoal.org).
[13] American Energy Alliance. Website. (Accessed 13 April 2021, as: http://www.americanenergyalliance.org).
[14] Coal News. Website. (Accessed 13 April 2021, as: http://www.coalnews.net).
[15] Coal Utilization Research Council. Website. (Accessed 13 April 2021, as: http://www.coal.org).
[16] IEA Clean Coal Centre. Website. (Accessed 13 April 2021, as: http://www.iea-coal.org.uk/).
[17] Lignite Energy Council. Website. (Accessed 13 April 2021, as: https://lignite.com/).
[18] Montana Coal Council. Website. (Accessed 13 April 2021, as: http://www.montanacoalcouncil.com/).
[19] New Mexico Mining Association. Website. (Accessed 13 April 2021, as: http://www.nmmining.org/).
[20] Texas Mining and Reclamation Association. Website. (Accessed 13 April 2021, as: http://www.tmra.com/).
[21] Utah Mining Association. Website. (Accessed 13 April 2021, as: http://www.utahmining.org/).

[22] Women in Mining. Website. (Accessed 13 April 2021, as: http://www.womeninmining.org/).

[23] Wyoming Mining Association. Website. (Accessed 13 April 2021, as: http://www.wyomingmining.org/).

[24] America's Power. Website. (Accessed 13 April 2021, as: http://www.americaspower.org).

[25] Arizona Mining Association. Website. (Accessed 13 April 2021, as: http://www.azcu.org).

[26] American Exploration & Mining Association. Website. (Accessed 13 April 2021, as: http://www.miningamerica.org/).

[27] National Coal Council. Website. (Accessed 13 April 2021, as: http://www.nationalcoalcouncil.org/).

[28] Peabody Energy. Website. (Accessed 13 April 2021, as: https://www.peabodyenergy.com/).

[29] Rocky Mountain Coal Mining Institute. Website. (Accessed 13 April 2021, as: www.rmcmi.org).

[30] American Coal Ash Association (ACAA). Website. (Accessed 13 April 2021, as: https://www.acaa-usa.org/).

[31] Ecoba European Coal Combustion Products Association. Website. (Accessed 30 April 2021, as: www.ecoba.com).

[32] Ash Development Association of Australia. Website. (Accessed 30 April 2021, as: www.adaa.asn.au).

[33] UK Quality Ash Association. Website. (Accessed 30 April 2021, as: www.ukqaa.org.uk).

[34] Association of Canadian Industries Recycling Coal Ash (CIRCA).

[35] A.R. Paris, A.B. Bocarsly. High-Efficiency Conversion of CO_2 to Oxalate in Water is Possible Using a Cr-Ga Oxide Electrocatalyst. ACS Catalysis. 2019, 9(3), 2324–2333.

[36] R. Angamuthu, P. Byers, A.L. Martin Lutz, E.B. Spek. Electrocatalytic CO_2 Conversion to Oxalate by a Copper Complex. Science. 327(5693), 313–315. Downloadable as https://science.sciencemag.org/content/5/5963/313.full.

[37] L.A.J. Garvie. Decay of cacti and carbon cycling. Naturwissenschaften. 2006, 93, 114–118.

[38] C. Gougoulias, J.M. Clark, L.J. Shaw. The role of soil microbes in the global carbon cycle: Tracking the below-ground microbial processing of plant-derived carbon for manipulating carbon dynamics in agricultural systems. Journal of the Science of Food and Agriculture. 2014 Downloadable at. https://www.ncbi.nlm.nih.gov/pmc/articles/PMC4283042/pdf/jsfa0094-2362.pdf.

[39] Global Monitoring Laboratory. Website. (Accessed 8 May 2021, as: https://Esrl.noaa.gov/gmd/ccgg/trends/).

Chapter 3
Oil and natural gas

3.1 Introduction

The use of oil and natural gas as a chemical fuel source does have an ancient history, but not in terms of anything that can be considered a large scale. In various locations, small amount of oil does seep up to the surface of the Earth and, in the past, has been gathered and burned. But the idea of distilling it into some usable fuel is a much more recent use for crude oil. Beyond that, the oil that is mentioned in ancient texts, such as the Bible, was routinely some material like olive oil, and not the oil we think of today, that produced from deposits underground.

The discovery of crude oil in Black Creek, Ontario, Canada, in 1858 and Titusville, Pennsylvania, in 1859 have gone down in history as a turning point in how oil was produced and used, most readily as a fuel for the kerosene lanterns that lit homes at the time [1, 2]. Since whale oil was the predominant source of such fuel for home lamps prior to the discovery of what was then called "rock oil," it is now believed that the discovery and ultimate shift in the source saved several species of whale. Predications are that had the new fuel not been found that humans would have hunted several species of whale to extinction.

Perhaps, obviously, the growth of the oil industry has been enormous and has involved governments and businesses in many countries, most notably the automobile industry. Such growth has been coupled with the formation of a large number of national and international organizations to promote its extraction, refining, and use [3–49]. In addition to this, a large number of regional organizations exist which are also involved with, or that advocate for, the production of petroleum or petroleum products [50–106]. Even some of the names of modern petroleum products harken back to the geographic origins of their sources, as shown in Figure 3.1 – the "Penn" connection being to Pennsylvania.

3.2 Geographic locations

The just-mentioned Titusville, Pennsylvania, is hardly the only place in which oil has been extracted from the ground profitably. In the United States, Texas is still considered oil country, as is Oklahoma and Alaska. But there are several other nations which are major producers of oil, and decades ago, on 14 September 1960, they formed an association that remains very powerful today, OPEC [3].

The Organization of Petroleum Exporting Countries (OPEC) accounts for slightly less than one half of the global production of oil, as of 2018.

https://doi.org/10.1515/9783110662276-003

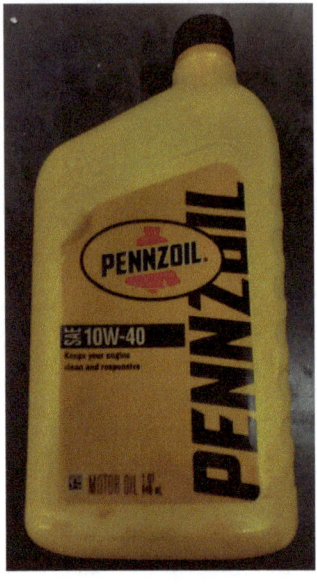

Figure 3.1: Commercially available motor oil.

Table 3.1: OPEC nations [3].

Nation	Date in OPEC	Proven reserves (M bbl)	Other export(s)	Comments
Algeria	1969	12,200	Fertilizers, ammonia	
Angola	2007	8,423	Diamonds	>90% of exports are petroleum
Ecuador	1973	8,273	Bananas, cacao, coffee	
Equatorial Guinea	2017	1,100	Woods, gold	Largely petroleum
Gabon	1975	2,000	Timber, uranium	~70% petroleum
Iran	Original	157,530	Chemicals	~50% petroleum, largely to China
Iraq	Original	143,069		>90% crude oil
Kuwait	Original	101,500	Fertilizer	
Libya	1962	48,363	Chemicals	
Nigeria	1971	37,070	Cocoa, aluminum	
The Republic of the Congo	2018	1,600	Copper, woods	

Table 3.1 (continued)

Nation	Date in OPEC	Proven reserves (M bbl)	Other export(s)	Comments
Saudi Arabia	Original	266,578		Generally, de facto leader of the cartel
United Arab Emirates	1967	97,800	Aluminum, other metals	
Venezuela	Original	299,953	Chemicals	

While the OPEC nations, shown in Table 3.1, produce a large amount of the world's crude oil, the oil industry is large enough, and international enough that there is a huge number of state, regional, national, or international organizations devoted to the production and marketing of oil and the products produced from it [2–107].

3.3 Sweet and sour crude

Much like metal ores are found in different locations in the world with differing levels of purity, crude oil is found in different areas of the world with different compositions of hydrocarbons, and with different amounts of other materials, which are considered pollutants because they lower the quality of the crude, and make purification a more complex process. These materials include nitrogen-containing compounds, sulfur-containing compounds, and mercury sulfides.

In terms of definitions, formally, sweet crude is one that has less than 0.5% sulfur in any form. Thus, there is less energy input in the removal of sulfur as the crude is processed.

Therefore, any crude oil batch or well that produces oil with more than 0.5% sulfur in it is defined as sour. Sour crudes are still quite usable and can be refined but require enhanced recovery in terms of sulfur and other pollutants, such as the just-mentioned mercury.

Failure to remove sulfur from any grade of crude oil leads to high levels of pollution, much of which has traditionally been released to the atmosphere. In the past few decades, sulfur has been removed from crude oil through a variety of treatments. For example, sodium plumbite – $Na_2PbO_2 \cdot 3H_2O$ – can be used to produce disulfides, thus removing the odor of sulfur compounds from the starting material, although not separating them from the mix. This is referred to as the doctor sweetening process.

More recently, what are called hydrodesulfurization (HDS) processes are used to remove the sulfur compounds from crude oil. Hydrogen sulfide is produced by mixing hydrogen with the batch, and using a cobalt–molybdenum catalyst, often supported on alumina. But other methods of sulfur removal exist, and this remains an active area of research. Figure 3.2 shows an example of the reaction.

$H_{2(g)} + RC_2H_4SH \rightarrow H_2S_{(g)} + RC_2H_5$ **Figure 3.2:** Hydrodesulfurization.

The "R" in Figure 3.2 represents any organic fragment, as it does in much of organic chemistry, since the reaction proceeds with a variety of sulfur-containing hydrocarbon molecules.

3.4 Natural gas

The term "natural gas" is used to describe the mixture of light-molecular-weight hydrocarbons – usually with methane as the gas present among these in the highest amount, and ethane following that – that naturally occur underground, that are used extensively as a hydrocarbon fuel source, and that are very often colocated with the heavier molecular weight hydrocarbons in crude oil. Their separation from crude oil is described further.

3.4.1 Natural gas and power generation

When used for power generation, natural gas serves as the hydrocarbon feed that ultimately turns turbines, which generates power, as discussed in Chapter 2. The most efficient plants are combined cycle gas turbine plants, since they combine gas turbines along with steam turbines.

3.4.2 Transportation and natural gas

Automobiles, buses, and trucks can be produced which run on natural gas. Natural gas has an octane number in the range of 120–130, which is significantly higher than the 87 of gasoline. Such engines have the advantage of cleaner burning fuel than traditional gasoline in internal combustion engines. However, massive infrastructure is already in place worldwide for the use of gasoline and diesel engines; and thus it appears that some major displacement by natural-gas-burning vehicles is not economically practical.

3.4.3 Household use of natural gas

Energy generation in the form of some household application is another method of using natural gas. Many residential areas have piping systems that bring natural gas to homes. Its use in central heating for homes is a major use, as is its use in stoves and ranges as a cooking fuel.

The domestic use of natural gas is one in which people are often aware that a very small amount of a specific pollutant is added intentionally. A small amount of H_2S – the rotten egg gas – is added to gas that is piped into homes, generally 1–10 ppm. This is still often simply called "mercaptan." It is a safety precaution. Natural gas is odorless, but should there be any leak in the system, the adulterant H_2S gas can be smelled easily, and the proper authorities notified.

3.4.4 Syngas

The production of carbon monoxide and hydrogen gas – what is called syngas – is another use of methane, which is natural gas, as shown in Figure 3.3. Figure 3.4 shows the further reaction of the carbon monoxide formed in Figure 3.3, called the water gas shift reaction, which produces carbon dioxide. Both reactions, however, are designed to produce hydrogen for further reaction with nitrogen to make ammonia. Most of this is used for fertilizer, and not for energy production.

$H_2O + CH_{4(g)} \rightarrow CO_{(g)} + H_{2(g)}$ **Figure 3.3:** Syngas production.

$CO_{(g)} + H_2O \rightarrow H_{2(g)} + CO_{2(g)}$ **Figure 3.4:** Water gas shift reaction.

Clearly, natural gas has important uses that are not involved in energy production. Since some of these are older and more established uses, any increased use of natural gas for the production of energy will have to compete with these needs.

3.4.5 Natural gas and fertilizer

Most of the fertilizers used in the world today come from the direct combination of nitrogen gas, isolated from air, and hydrogen gas, produced as shown earlier. This is called the Haber process and is shown in Figure 3.5. This reaction has in the past

$3H_{2(g)} + N_{2(g)} \rightarrow 2NH_{3(g)}$ **Figure 3.5:** The Haber process.

utilized hydrogen extracted from coal but now tends to use hydrogen produced as shown earlier, in syngas and the water gas shift reactions.

Natural gas has other uses as well, but its uses for power generation in power plants, buildings, and vehicles are three of its largest.

3.5 Refining

Refining crude oil into several useful classes of organic molecules has become a multi-trillion dollar industry, and can be practiced at the site at which crude oil is refined or at sites far distant from the point of extraction of the crude. Crude oil batches differ in composition depending on their global location, and thus, the specific steps of any manner of what is called a refining operation can certainly change according to an individual feedstock or an individual batch. Yet several very broad steps exist which all refineries must incorporate and in some way use. Often they can be defined and categorized through use of the temperature range required for each step. These include the following:

3.5.1 Desalting

Numerous types and varying amounts of suspended materials do exist in different batches of crude oil. The desalting step removes materials, including suspended sand, salts, and clays, routinely in a temperature range of 60–90 °C. Additionally, desalting can also separate out components of fracking fluids, those that are comingled into the crude oil during the hydraulic fracturing process.

3.5.2 Distillation

Crude oil distillation generally occurs at approximately 400 °C, often at somewhat elevated pressure. This early-stage process is designed to begin separating compounds (there are thousands in crude oil) into fractions that are close in boiling point, and thus often close to the molecular weight.

3.5.3 Hydrotreating or hydroprocessing

Heavier hydrocarbons must be broken down to smaller molecular weight compounds, with one of the target weights being those molecules that fall into C8 fraction. Elevated pressures and temperatures, generally 200–300 psi and 350–400 °C, are required. At this point, elemental hydrogen is added, which aids in the transformation of lighter molecular weight hydrocarbon molecules. Oftentimes, it is methane that serves as the hydrogen source (the H_2). The lighter fraction which has been separated serves as the source, since it can readily be stripped of its hydrogen atoms.

3.5.4 Cracking or hydrocracking

This is a further refining step that is aimed at improving the amount of octane or C8 fraction of any hydrocarbon feedstock. The term "cracking" implies that larger, heavier hydrocarbon molecules are being broken down into lighter ones. Longer contact times are required for these chemical transformations, and elevated temperatures and pressures are still required. Both of these broad steps aim to increase the amount of molecules that can be used as motor fuels.

3.5.5 Coking

Another step in the distillation process that is used to break larger molecules to smaller ones is coking. Run at approximately 450 °C, this is another means by which the amount of motor fuel, the C8 fraction, is enhanced. At times called destructive distillation, coking is a means of severe thermal cracking.

3.5.6 Visbreaking

Yet another step called visbreaking takes place at approximately 480 °C, which was designed to take lower value, higher molecular weight hydrocarbons and break them down to higher value, lower molecular weight molecules. Since fuels such as diesel fuel, gasoline, or heating oil represent higher value materials, these again become the target molecules in such a transformation. The term "visbreaking" is related to viscosity because the process runs at a high enough temperature that the mixture is far less viscous than at ambient temperatures. It is essentially another form of destructive distillation.

3.5.7 Steam cracking

Steam cracking is another high-temperature process. This step produces olefins (alkenes) and must take place at approximately 850 °C. At such temperature, steam cracking works with a wide variety of hydrocarbon feedstocks, from something as low in molecular weight as ethane to much higher molecular weight hydrocarbons. The process temperature is high enough that it breaks down higher molecular weight alkanes at the same time in which it forms alkenes. Product output can be affected by the catalyst used in such an operation. Very often, this step is the main means by which ethylene and propylene are produced from feedstocks such as liquefied petroleum gas or naphtha.

3.5.8 Catalytic reformers

Catalytic reforming is run at approximately 430–500 °C, with the aim of producing hydrocarbons with high branching, and molecular weights near that of octane. Naphtha is often the feedstock, and the product is referred to as a reformate. This process also produces cyclic compounds that are dehydrogenated into aromatics. Both are aimed once again to enhance the amount of motor fuel material in the target or product. This step also produces a significant amount of elemental hydrogen, which can be used in other steps within the overall refining process. Benzene, toluene, and xylene (often called BTX) are also produced in this step.

3.5.9 Alkylation

Highly branched hydrocarbon molecules with a total of eight carbon atoms – often just called the C8 fraction – are a desired target, because they burn most efficiently in internal combustion engines. Thus, the alkylation step reacts lower molecular weight olefins with paraffins or paraffin-like molecules with the end result being highly branched alkane hydrocarbon molecules. In this set of reactions, isobutene is mainly used for the manufacture of various liquid fuels, as shown in Figure 3.6. The reaction does require some catalysts – sulfuric acid or hydrofluoric acid have been found to work well – to initiate the transformation. The resulting material becomes an important molecule in the mixture, that is, motor fuel. It improves the overall octane number of motor gasoline, seen by the consumer at the pump.

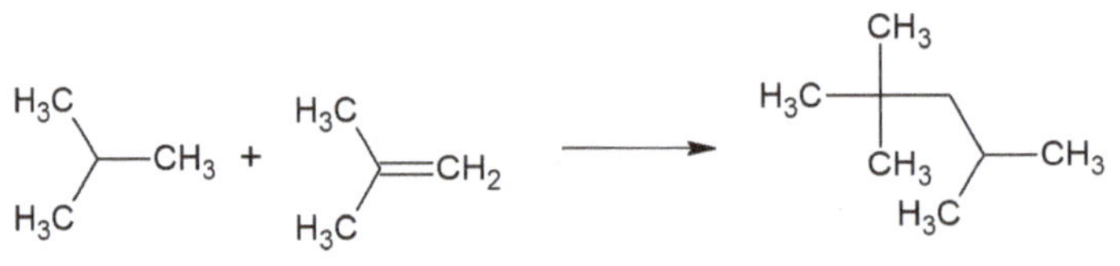

Figure 3.6: Lightweight olefin alkylation.

The production of 2,2,4-trimethylpentane is particularly important, because it is the component of C8 fraction that burns the best in automobile engines. The term "octane" is used for this fraction and this mixture, but *n*-octane actually does not burn particularly well in automobile engines.

3.5.10 Isolation of C1, or natural gas fraction

The major component of natural gas by volume is methane (CH_4), which is separated along with other light hydrocarbons, all of which are components of low-molecular-weight gas fraction. If separation can be done by distillation, then the components of the mixed gas further separated into methane, ethane, propane, butane, and other gaseous molecules. As well, such materials can be transported without further separation, if needed, as shown in Figure 3.7.

Methane is a valuable small hydrocarbon because it can be stripped of its hydrogen atoms so that hydrogen can be subsequently utilized in other chemical processes, including hydrotreating, in which sulfur can be removed from the crude mix.

Figure 3.7: Transport of liquefied petroleum gas.

3.5.11 Recovery of sulfur by-products

Sulfur oxides (generally called SO_x or SOX) must be removed from crude oil during refining so that such compounds do not become pollutants that escape to the atmosphere. Such effluents have done significant environmental damage in some areas in the past. One method of removal of SOX is the capture of it as hydrogen sulfide (H_2S), which can subsequently be reacted and treated using oxygen, the end result being the production of sulfuric acid (H_2SO_4), the largest chemical commodity produced globally, from several different sources.

3.6 Transportation fuel use

After fractionation, there are many uses for the refined products of crude oil. One that is now obvious to virtually everyone is the fraction that is used for automobile gasoline, the octane fraction, as shown in the picture of a gasoline pump, in Figure 3.8. Notice that different numbers are assigned to different levels of gasoline, based on their composition.

Figure 3.8: Gas station and gasoline fractions.

But there are other types of transportation fuel as well, since all internal combustion engines do not run on gasoline. Larger engines, or those with different performance specifications, require different fuels, such as diesel, jet fuel, or various marine fuels. Table 3.2 shows this based on the ranges of molecular weights.

Table 3.2: Fuel types.

Name	Carbon number	BP (°C)
Gasoline	C_8	~200
Naphtha	C_3–C_{10}	40–175
Kerosene	C_{10}–C_{13}	175–250
Diesel	C_{13}–C_{20}	240–350
Light gas oil	C_{20}–C_{30}	350–450
Heavy gas oil	C_{30}–C_{44}	450–540

There are higher molecular weight fractions as well, but those listed in Table 3.2 are those most used for energy production in terms of transportation. Figure 3.9 shows both the Lewis structure for *n*-octane, as well as that for 2,2,4-trimethylpentane. Both are isomers of C_8H_{18}, but the branched isomer combusts much better in the engines of most automobiles.

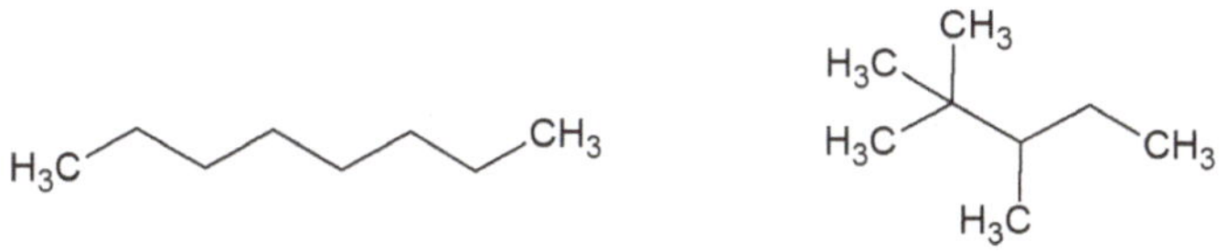

Figure 3.9: C-8 fraction isomers.

3.7 Nonfuel use and by-products

Again it may seem obvious that not all the products of crude oil distilling and refining are used for the production of energy. The monomers of commodity plastics are all largely extracted and purified from crude oil, although there are growing attempts to meet at least some of the needs for such monomers from renewable, plant-based sources.

We will not engage in a full discussion of the production of plastic monomers but will say that monomer and fuel are the two major uses of crude oil, and the products into which it is refined. The most common plastics are shown in Table 3.3. There are other uses for the heavier fractions as well, essentially so that they do not become waste products. For example, the heaviest fraction of crude oil, called tar, or natural bitumen, cannot easily be broken down into lighter molecular weight molecules at any heat that is economically feasible to generate. Thus, a significant amount of it is used in road construction and in other similar applications.

Table 3.3: Most common plastics.

Name	RIC	Lewis structure, repeat unit	Example uses	Comments
Polyethylene-terephthalate	1 (PETE)		Beverage bottles	Often recycled
High density polyethylene	2 (HPDE)		Plastic lumber, plastic bags	Often recycled
Polyvinyl chloride	3 (PVC or V)		Piping, household components	
Low-density polyethylene	4 (LDPE)		Containers, six-pack rings	Seldom recycled
Polypropylene	5 (PP)		Food containers and lids	Recycled, depending on locale
Polystyrene	6 (PS)		Packing material	Not usually recycled when formed as Styrofoam ®
Other	7	Various	Very wide variety of uses	Not normally recycled

3.8 Recycling plastics for fuel for power generation

Having just mentioned the production of plastics from crude oil sources, it is also worth looking at their recycling, and examining a use that can be their end use – this latter use is important in energy generation.

The recycling of plastics from the consumer household or other end user is now a common, mature industry in many nations. Since the United States is a geographically large, diverse country, with areas of very low population density, there remains no national policy about the recycling of plastics, as opposed to their one-time use and disposal. Rather, each state makes laws concerning this, under the direction of their governor and state house and senate. Other countries, some with more centralized governments, have been able to establish national policies for recycling and reuse. Table 3.4 shows again what are called RICs – resin identification codes – that are in use for plastics that are produced in the greatest amounts.

Table 3.4: Plastic RICs.

RIC	Name	Abbreviation	Comments
1	Polyethylene terephthalate	PETE, or PET	Recycled in large amounts
2	High-density polyethylene	HDPE, or PE-HD	
3	Polyvinyl chloride	PVC, or V	Not routinely recycled to fuel, chlorine presents problem
4	Low-density polyethylene	LDPE, or PE-LD	
5	Polypropylene	PP	Can be converted to fuel; reacted in water at 400–500 °C.
6	Polystyrene	PS	Seldom recycled, as Styrofoam®
7	Other	O	Seldom recycled

The recycling of PETE is very common, often with the end result being the production of new consumer objects – like beverage bottles – from plastic items that had already been used for such consumer products. But other high-volume plastics are recycled as well, and their RIC codes are seen on an enormous number of end user items. Figure 3.10 shows just one example of a product that is recyclable, and that is an end user item.

Some plastics are not easy to recycle, however, based on the cost involved in transporting them from a plastics recycling collection center to a firm that can again make them into some usable product. Polystyrene is an excellent example because it is often made into some low-density material, such as protective packaging (often called Styrofoam® by the general public). Thus, alternate uses for these polymers are sought after they have been used once, especially if they have in some way been degraded, for example, feedstock in incinerators.

In numerous countries, incineration of once-used plastics for fuel for power generation is not unheard of but, in some locations, faces intense popular opposition. The incinerators that are able to burn plastics cleanly tend to be more expensive to construct than what are called traditional incinerators. The public opposition is usually based on the performance of older, traditional incinerators, which do not burn plastics to completion, and produce unpleasant, noxious odors that have been linked to various diseases in humans.

Figure 3.10: Recyclable plastic food packaging lid.

One nation that handles incineration of waste material very well, including plastics, is Japan. The Japanese incineration industry is extremely careful about what material goes into any incinerator, segregates plastics from other feedstock materials, and burns different materials at different temperatures, ensuring the best possible combustion of all starting materials. Outside observers and visitors have claimed in the popular press that Japanese incinerator operations appear to an outsider to be as clean as a food processing plant.

Since some incinerators link their heat production to energy production, this then becomes a viable form of energy generation, even if it is small overall, when compared to the use of coal and oil for energy and electricity production.

Another route by which plastics can be turned into some form of fuel is a high-temperature pyrolysis. Heating the material to a high enough temperature, often under enhanced pressure, can break down the plastic, not necessarily to its starting monomers, but to small enough molecules, resulting in some liquid fuel. The general steps in such a process are as follows:

1. Bulk plastic waste
2. Cleaning
3. Crushing to small pieces
4. Pyrolysis
5. Reclamation of fuel – usually diesel [107]

This type of subsequent use of plastics remains a small one when compared to the recycling programs that are already in place worldwide.

3.9 The politics of oil and natural gas

No other commodity has shaped the political landscape as has the production, sale, and use of oil and natural gas, and the motor fuels and plastics made from them. The end of the Second World War – the first truly mechanized war with multitudes of vehicles that ran on gasoline or diesel – and the rise of what can be called the oil culture appear to have both occurred at the same general time. From this, and because developed nations were taking oil from several underdeveloped nations at very favorable prices, the OPEC was formed. We have listed the member nations already, in Table 3.1.

Among the OPEC nations, Saudi Arabia has traditionally been the member with the greatest influence. This is not a matter of chemistry such as whether their crude is sweeter or more sour but much more a matter of political muscle or pull. Saudi Arabia has a larger production capacity than the other OPEC nations, and thus has a significant ability to influence the price of crude oil on the world's markets, based on how much it chooses to produce and to sell.

Curiously, although the price of a barrel of crude oil is often agreed upon by the OPEC member nations, there have been times at which one or more member states violate the agreement, usually selling their oil at a lower, more favorable price than other member states. Such behavior usually results in some sort of price war and can affect the economies of both the producing nations, as well as those purchasing the oil.

While most adults in the developed world today know of a lifelong history of the OPEC nations exerting political leverage by adjusting the rate at which oil is released to world markets for refining, simply by adjusting the price of oil, the year 2020 saw what might be considered a reversal. Since the COVID-19 virus shutdown in huge parts of the world in 2020, billions of people suddenly stopped driving nearly as much as they normally did. This caused the price of gasoline at the pump to plummet. Figure 3.11 shows the prices at a gas station near Detroit, MI, USA, displaying prices for gasoline, in dollars per gallon, that are the lowest they have been in 25 years.

Gasoline prices, which in the United States had risen to higher than $4 per gallon at the pump in part of the 2010–2020 years, plunged in early 2020 to prices that had not been seen since the 1990s. This in turn meant that the price for a barrel of oil also dropped significantly. And this drop meant that the OPEC nations were actually losing money by exporting oil. Such price reversals are not common; and this one was the largest in living memory. When it actually costs to have crude oil unloaded from cargo ships, the term used is "negative price point." Such a situation has not been seen in the professional lifetime of most of those employed in the petrochemical industry.

This reversal in the cost of crude oil, and in the cost of motor fuels, has reinitiated a larger discussion of what is sometimes called a "post-oil future." Because it is recognized that the supply of oil will eventually run out, numerous ideas have been put forward about how to maintain the current quality of life without such dependence on oil, and how to diversify means of transportation. While a full list of possibilities is beyond the scope of this chapter, the following have become important

Figure 3.11: Gasoline prices, summer 2020.

aspects of the greater idea of moving people in cities, in greater than urban areas, and in wider, often rural areas.

List of alternative forms of urban transportation:

1. Inter-city light rail – cities such as Washington, DC, plus Alexandria, Virginia, and Frankfurt, Germany, have utilized this for decades, although there are a multitude of other such examples in the United States, Europe, Australia, Japan, and China as well.
2. Intra-city light rail – Utrecht, the Netherlands, has employed such a rail system since the 1970s.
3. Community owned, rentable bicycles – an idea that is growing worldwide, Denmark has used the Bycykler København bicycles for years. Recently, Detroit and surrounding suburbs in Michigan, USA, have adopted this idea, as shown in Figure 3.12.
4. Dedicated bicycle lanes. Many cities have designated certain parts of roads as reserved for bicycles, which can be either personal or community owned.
5. Carpooling lanes. Areas such as greater Washington, DC, have used high occupancy vehicle lanes as an incentive for people to carpool, save fuel, and reduce the number of cars on the road.
6. Electric, city trains – several European cities utilize electric trains. The idea is growing in the United States and Canada as well.

7. Buses. Well-established alternative buses with internal combustion engines or with electric motors have been used in many cities for decades. Some cities have in the recent past opted to expand their bus fleet and offer more options for users because they use less fuel and energy per person, and because they help relieve traffic congestion.
8. Alternative fuel vehicles, including electric vehicles. Vehicles that run on propane, hydrogen, or electric batteries all ease the stress on the amount of traditional gasoline and diesel fuel consumed.

Figure 3.12: Community bicycle stand.

Undoubtedly, in the future, the options adopted or encouraged for transportation by any city, state, or region will depend in part on the continued availability of traditional fuels but also on the economic aspects and feasibility of other, often nontraditional means such as those just mentioned. These means seem to be increasing with the passage of time.

References

[1] Development of the Pennsylvania Oil Industry. Website. (Accessed 20 April 2021, as: http://acs.org/content/acs/en/education/whatischemistry/landmarks/pennsylvaniaoilindustry.html).
[2] Drilling into Canada's Petroleum History. Website. (Accessed 20 April 2021, as: http://chemist.ca/magazine/article/drilling-into-canadas-petroleum-history).
[3] Organization of Petroleum Exporting Countries (OPEC). Website. (Accessed 22 September 2020, as: http://www.opec.org/opec_web/en/index.htm).

[4]   American Association of Petroleum Geologists (AAPG). Website. (Accessed 22 September 2020, as: http://www.aapg.org/).

[5]   American Association of Professional Landmen (AAPL). Website. (Accessed 22 September 2020, as: http://www.landman.org/).

[6]   American Gas Association (AGA). Website. (Accessed 22 September 2020, as: https://www.aga.org/).

[7]   American Exploration & Production Council (AXPC). Website. (Accessed 22 September 2020, as: http://www.axpc.us/).

[8]   American Institute of Chemical Engineers (AIChE). Website. (Accessed 22 September 2020, as: https://www.aiche.org/).

[9]   American Natural Gas Alliance (ANGA). Website. (Accessed 22 September 2020, as: https://www.naturalgassolution.org/).

[10]   American Petroleum Institute (API). Website. (Accessed 22 September 2020, as: http://www.api.org/).

[11]   American Public Gas Association (APGA). Website. (Accessed 22 September 2020, as: http://www.apga.org/home).

[12]   Association of American State Geologists (AASG). Website. (Accessed 22 September 2020, as: http://www.stategeologists.org/).

[13]   Association of Energy Engineers (AEE). Website. (Accessed 22 September 2020, as: https://www.aeecenter.org/).

[14]   Association of Energy Service Companies (AESC). Website. (Accessed 22 September 2020, as: http://www.aesc.net/).

[15]   Association of Energy Services Professionals International (AESP). Website. (Accessed 22 September 2020, as: http://www.aesp.org/).

[16]   Association of Professional Energy Consultants (APEC). Website. (Accessed 22 September 2020, as: http://apecmidwest.com/).

[17]   Australian Gas Association. Website. (Accessed 22 September 2020, as: www.aga.asn.au).

[18]   Australian Petroleum Production & Exploration Association (APPEA). Website. (Accessed 22 September 2020, as: www.appea.com.au).

[19]   Canadian Association of Petroleum Producers (CAPP). Website. (Accessed 22 September 2020, as: https://www.capp.ca/).

[20]   Canadian Energy Pipeline Association (CEPA). Website. (Accessed 22 September 2020, as: https://cepa.com/en/).

[21]   Canadian Gas Association (CGA). Website. (Accessed 22 September 2020, as: http://www.cga.ca/).

[22]   Energy Bar Association (EBA). Website. (Accessed 22 September 2020, as: http://www.eba-net.org/).

[23]   European Federation of Energy Traders (EFET). Website. (Accessed 22 September 2020, as: http://www.efet.org/).

[24]   Fuels Europe. Website. (Accessed 22 September 2020, as: fuelseurope.eu).

[25]   Gas Producers Association (GPA). Website. (Accessed 22 September 2020, as: https://www.aga.org/).

[26]   Gas Research Institute (GRI)/ Gas Technology Institute (GTI). Website. (Accessed 22 September 2020, as: http://www.gastechnology.org/About/Pages/History.aspx).

[27]   Geological Society of America (GSA). Website. (Accessed 22 September 2020, as: https://www.geosociety.org/).

[28]   Independent Petroleum Association of America. Website. (Accessed 22 September 2020, as: www.ipaa.org).

[29] National Association of Energy Service Companies (NAESCO). Website. (Accessed 22 September 2020, as: http://www.naesco.org/).

[30] National Association of Royalty Owners (NARO). Website. (Accessed 22 September 2020, as: http://www.naro-us.org/).

[31] National Energy Marketers Association (NEM). Website. (Accessed 22 September 2020, as: https://www.energymarketers.com/).

[32] National Energy Services Association (NESA). Website. (Accessed 22 September 2020, as: http://www.nndb.com/org/520/000126142/).

[33] National Petroleum Council (NPC). Website. (Accessed 22 September 2020, as: http://www.npc.org/).

[34] National Propane Gas Association (NPGA). Website. (Accessed 22 September 2020, as: https://www.npga.org/).

[35] National Stripper Well Association (NSWA). Website. (Accessed 22 September 2020, as: http://nswa.us/custom/showpage.php?id=15).

[36] Natural Gas Supply Association (NGSA). Website. (Accessed 22 September 2020, as: http://www.ngsa.org/).

[37] New Zealand Gas Industry, GasNZ. Website. (Accessed 22 September 2020, as: http://gasnz.org.nz/nz-gas-industry).

[38] North American Energy Standards Board (NAESB). Website. (Accessed 22 September 2020, as: https://www.naesb.org/).

[39] Petroleum Exploration & Production New Zealand (PEPANZ). Website. (Accessed 22 September 2020, as: www.pepanz.com).

[40] Petroleum Marketers Association of America (PMAA). Website. (Accessed 22 September 2020, as: http://www.pmaa.org/).

[41] Society of Independent Professional Earth Scientists (SIPES). Website. (Accessed 22 September 2020, as: https://sipes.org/).

[42] Society of Petroleum Engineers (SPE). Website. (Accessed 22 September 2020, as: http://www.spe.org/unitedstates/).

[43] Society of Petroleum Evaluation Engineers (SPEE). Website. (Accessed 22 September 2020, as: https://secure.spee.org/).

[44] Society of Professional Well Log Analysts (SPWLA). Website. (Accessed 22 September 2020, as: https://www.spwla.org/).

[45] South Africa Oil & Gas Alliance (SAOGA). Website. (Accessed 22 September 2020, as: www.saoga.org.za).

[46] South African Gas Association. Website. (Accessed 22 September 2020, as: http://saqccgas.co.za/associations/sapga).

[47] South African Petroleum Industry Association. Website. (Accessed 22 September 2020, as: www.sapia.org.za).

[48] United States Energy Association (USEA). Website. (Accessed 22 September 2020, as: https://www.usea.org/).

[49] U.S. Oil & Gas Association (USOGA). Website. (Accessed 22 September 2020, as: https://www.usoga.org/).

[50] Alabama Natural Gas Association (ANGA). Website. (Accessed 22 September 2020, as: http://alnga.org/).

[51] Alaska Oil & Gas Association (AOGA). Website. (Accessed 22 September 2020, as: https://www.aoga.org/).

[52] Arkansas Independent Producers Royalty Owners Association (AIPRO). Website. (Accessed 22 September 2020, as: http://aipro.org/).

[53] Texas Pipeline Association (TPA). Website. (Accessed 22 September 2020, as: http://www. texaspipelines.com/).

[54] California Independent Petroleum Association (CIPA). Website. (Accessed 22 September 2020, as: https://www.cipa.org/i4a/pages/index.cfm?pageid=1).

[55] Colorado Oil & Gas Association (COGA). Website. (Accessed 22 September 2020, as: http:// www.coga.org/).

[56] Eastern Kansas Oil & Gas Association (EKOGA). Website. (Accessed 22 September 2020, as: http://www.ekoga.org/).

[57] Energy Association of Pennsylvania. Website. (Accessed 22 September 2020, as: http:// www.energypa.org/).

[58] Florida Independent Petroleum Producers Association Inc. (FLIPPA). Website. (Accessed 22 September 2020, as: http://www.flippaoil.org/).

[59] Florida Natural Gas Association (FNGA). Website. (Accessed 22 September 2020, as: http:// floridagas.org/).

[60] Houston Energy Club. Website. (Accessed 22 September 2020, as: http://www.houstonener gyclub.org/).

[61] Illinois Oil & Gas Association (IOGA). Website. (Accessed 22 September 2020, as: http:// www.ioga.com/).

[62] Independent Oil and Gas Association of New York (IOGANY). Website. (Accessed 22 September 2020, as: http://www.iogany.org/).

[63] Independent Oil & Gas Association of Pennsylvania. Website. (Accessed 22 September 2020, as: https://www.pioga.org/).

[64] Independent Oil & Gas Association of West Virginia (IOGA-WV). Website. (Accessed 22 September 2020, as: https://iogawv.com/).

[65] Louisiana Oil & Gas Association (LOGA). Website. (Accessed 22 September 2020, as: http:// loga.la/).

[66] Louisiana Mid-Continent Oil and Gas Association. Website. (Accessed 22 September 2020, as: http://www.lmoga.com/).

[67] Michigan Basin Geological Society. Website. (Accessed 22 September 2020, as: https:// www.mbgs.org/).

[68] Michigan Oil and Gas Association (MOGA). Website. (Accessed 22 September 2020, as: http://www.michiganoilandgas.org/).

[69] Midwest Energy Association (MEA). Website. (Accessed 22 September 2020, as: https:// www.guidestar.org/profile/41-0855255).

[70] Midwest Cogeneration Association (MCA). Website. (Accessed 22 September 2020, as: http://www.cogeneration.org/).

[71] Mississippi Independent Producers and Royalty Owners (MIPRO). Website. (Accessed 22 September 2020, as: http://www.mipro.ms/).

[72] Montana Petroleum Association (MPA). Website. (Accessed 22 September 2020, as: https:// montanapetroleum.org/).

[73] Mountain States Legal Foundation (MSLF). Website. (Accessed 22 September 2020, as: https://www.mountainstateslegal.org/).

[74] University of Houston Energy Association (HEA). Website. (Accessed 22 September 2020, as: http://www.uhenergyassociation.org/).

[75] Natural Gas & Energy Association of Oklahoma (NGEAO). Website. (Accessed 22 September 2020, as: http://www.ngeao.org/).

[76] Natural Gas Society of East Texas (NGSET). Website. (Accessed 22 September 2020, as: http://www.ngset.org/).

[77] Natural Gas Society of the Permian (NGSPB). Website. (Accessed 22 September 2020, as: https://ngspb.com/).

[78] New Mexico Oil & Gas Association (NMOGA). Website. (Accessed 22 September 2020, as: https://www.nmoga.org/).

[79] New Orleans Geological Society (NOGS). Website. (Accessed 22 September 2020, as: http://www.nogs.org/).

[80] New York State Oil Producers Association (NYSOPA). Website. (Accessed 22 September 2020, as: http://www.newyorkstateoilproducersassociation.com/).

[81] North Dakota Petroleum Council. Website. (Accessed 22 September 2020, as: https://www.ndoil.org/).

[82] Northeast Gas Association (NGA). Website. (Accessed 22 September 2020, as: http://www.northeastgas.org/index.php).

[83] Northwest Gas Association (NWGA). Website. (Accessed 22 September 2020, as: https://www.nwga.org/).

[84] Ohio Oil & Gas Association (OOGA). Website. (Accessed 22 September 2020, as:http://www.ooga.org/).

[85] Oklahoma Independent Petroleum Association (OIPA). Website. (Accessed 22 September 2020, as: http://www.oipa.com/custom/index.php).

[86] Oklahoma Oil and Gas Association. Website. (Accessed 22 September 2020, as: http://okoga.com/).

[87] Panhandle Producers & Royalty Owners Association (PPROA). Website. (Accessed 22 September 2020, as: https://pproa.org/).

[88] Pennsylvania Independent Oil and Gas Association (PIOGA). Website. (Accessed 22 September 2020, as: https://www.pioga.org/).

[89] Permian Basin Geophysical Society (PBGS). Website. (Accessed 22 September 2020, as: http://pbgsmidland.org/).

[90] Permian Basin Landmen's Association (PBLA). Website. (Accessed 22 September 2020, as: http://www.pbla.org/about/).

[91] Permian Basin Petroleum Association (PBPA). Website. (Accessed 22 September 2020, as: http://pbpa.info/).

[92] Permian Basin Society of Petroleum Engineers (SPE-PB). Website. (Accessed 22 September 2020, as: http://connect.spe.org/permianbasin/home).

[93] Petroleum Association of Wyoming (PAW). Website. (Accessed 22 September 2020, as: http://www.pawyo.org/).

[94] Rocky Mountain Mineral Law Foundation (RMMLF). Website. (Accessed 22 September 2020, as: https://www.rmmlf.org/).

[95] South Texas Geological Society (STGS). Website. (Accessed 22 September 2020, as: http://www.stgs.org/).

[96] Southeastern Gas Association (SGA). Website. (Accessed 22 September 2020, as: https://www.southerngas.org/).

[97] Southwest Kansas Royalty Owners Association (SWKROA). Website. (Accessed 22 September 2020, as: http://www.swkroa.com/).

[98] Tennessee Oil & Gas Association (TOGA). Website. (Accessed 22 September 2020, as: http://www.tennoil.com/).

[99] Texas Alliance of Energy Producers. Website. (Accessed 22 September 2020, as: http://texasalliance.org/).

[100] Texas Independent Producers & Royalty Owners Association (TIPRO). Website. (Accessed 22 September 2020, as: https://www.tipro.org/).

[101] Texas Oil & Gas Association (TXOGA). Website. (Accessed 22 September 2020, as: https://www.txoga.org/about-us/).

[102] Texas Pipeline Association (TPA). Website. (Accessed 22 September 2020, as: http://www.texaspipelines.com/).

[103] Texas Section of the American Institute of Professional Geologists (AIPG). Website. (Accessed 22 September 2020, as: http://aipg-tx.org/).

[104] West Central Texas Oil & Gas Association. Website. (Accessed 22 September 2020, as: https://www.txoga.org/about-us/).

[105] Western Energy Alliance (WEA). Website. (Accessed 22 September 2020, as: https://www.westernenergyalliance.org/).

[106] Western States Petroleum Association (WSPA). Website. (Accessed 22 September 2020, as: https://www.wspa.org/).

[107] Anthropocene Magazine. Website. (Accessed 5 June 2021, as: www.anthropocenemagazine.org).

Chapter 4
Biofuels

4.1 Introduction – bioethanol

4.1.1 Starch sources

The use of plant sources to produce ethanol has an ancient history and is one of the earliest recorded forms of a chemical reaction monitored by people. One can make the claim that the rise of civilizations has been accompanied by the fermentation of grains and grapes, which always resulted in a variety of beers and wines [1]. Throughout almost all of this history, the consumption of alcoholic beverages was not simply for pleasure and an intoxicating effect; it was for health and safety. Millennia before any understanding of how germs in water could make people ill, it was known that mixing wine or beer with water in some way made the water safe to drink. As well, diluting alcoholic beverages with water made them easier for children to consume, in that they would not become inebriated. Perhaps the last vestige of this in our modern culture is the mixing of water and wine during Christian Masses and services, performed by the priest or minister, usually with some statement that it represents a metaphorical mixing of the humanity and divinity of Jesus.

While the roots of the history of fermentation may appear to be rather haphazard, different nations and cultures considered and regulated the production of beer and wine very seriously. Arguably the oldest written document concerning the purity of beer is the German Reinheitsgebot, or purity law. It states:

> Especially we want that forth in all our cities, markets and in the countryside to beer no more pieces than barley, hops and water should be used.

Note that this famous law does not include any mention of yeast, the biocatalyst that initiates fermentation. This was simply because the law, dated 1516, was put into effect before any understanding of microorganisms such as yeast existed.

Until the incorporation of carbon-based molecules from crude oil was used for ethanol production, all ethanol that was produced was bioethanol, and all was produced using the rather simple chemistry shown in Figure 4.1.

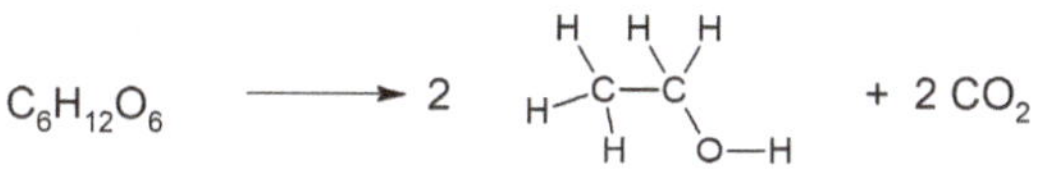

$$C_6H_{12}O_6 \longrightarrow 2 \; CH_3CH_2OH + 2\,CO_2$$

Figure 4.1: Bioethanol production.

https://doi.org/10.1515/9783110662276-004

It is difficult to represent yeast in the reaction, since its role is essentially a catalytic one and the yeast is a microorganism. Note also that for every mole of the sugar that is converted into 2 mol of ethanol, 2 mol of carbon dioxide are coproduced as a by-product. This has traditionally been responsible for the bubbles in beers and wines.

Beers and wines produced from grains and grapes are not the only means by which alcoholic beverages have been produced traditionally. Table 4.1 lists several other sources of ethanol – and is not a complete list, as several other plants have been used on a localized, relatively small scale.

Table 4.1: Alcohol production from plants.

Plant source	Beverage name	Traditional area
Apples	Cider	Europe, North America
Barley	Beer	Europe
Corn	Bourbon	North America
Grapes	Wine, cognac, brandy	Wines – worldwide
Rye	Rye beer	Europe
Millet	Millet beer	Eastern Asia
Wheat	Wheat beer	Several areas, originally the Near East
Sorghum	Sorghum beer	Western Africa
Juniper berries	Gin	Europe
Honey	Mead	Northern Europe
Rice	Saki	Japan
Rice	Sonti	India
Sugar cane	Rum	Caribbean
Agave	Tequila	Central America
Potato	Vodka	Northern, eastern Europe

These products continue to be produced through the fermentation of plants, but have largely become worldwide in their consumption. Our table is decidedly lacking in many of the alcoholic beverages that are called "spirits," since such beverages were not produced until people developed the ability to distill beers and wines, and thus increase the amount of alcohol in a given volume of any liquid. The production of scotch, as just one example, requires a still to re-condense the alcoholic portion of the starting liquid, and in the process concentrate the alcohol. After the distilling process, a great deal of effort and care goes into the storage of any resulting spirits, often in specific types of barrels at carefully controlled temperature and humidity conditions.

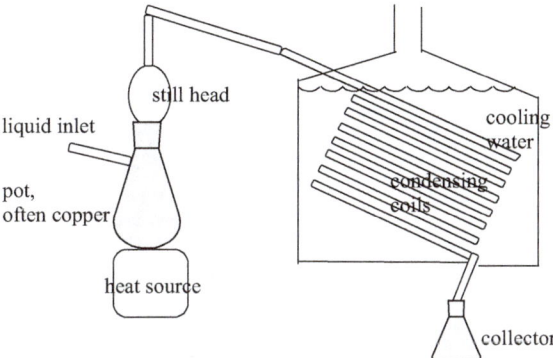

Figure 4.2: Distilling apparatus.

The distilling apparatus shown in Figure 4.2 is on purpose not shown to any particular scale. Even today, artisanal stills can be made quite small, for the home distiller who wishes to make spirits for personal or local use but can also be scaled up. Large stills must be used to concentrate the volumes of alcohol that are needed for motor fuels. Even the addition of 5% alcohol to motor gasoline, as occurs in much of the United States, requires an enormous amount of ethanol be distilled from sources that are allowed to ferment using various yeasts.

The early twentieth century was the first time in history that ethanol was made in some other way besides the fermentation of some plant matter. The use of petroleum distillates from the C2 fraction – the fraction of distillate with two carbon atoms – was found to be a cost-effective way to make ethanol. But this is counter to the current desire to produce ethanol for fuel from renewable sources. Such operations have become large enough though that there are now organizations that are either dedicated to the production of bioethanol, or that are deeply invested and involved in it [2–21].

4.1.2 Cellulosic ethanol

The production of cellulosic ethanol – ethanol made from the cellulose and not starch of plants – has become the aim of several researchers and companies in the past three decades. This and other non-starch sources of ethanol are referred to as second-generation sources, with any starch-based ethanol being first generation. The potential is great for a new, nonfossil source of ethanol, but the challenges also remain great. Often, cellulose is used for other products. In some areas, cellulose is plowed back into the soil, or left atop it so that wind does not erode the valuable topsoil and a thin topsoil layer. Chemically, cellulose is not degraded by yeast as is starch. The difference is the alpha versus the beta linkage in the sugar rings of each material. Figure 4.3 illustrates the two isomers.

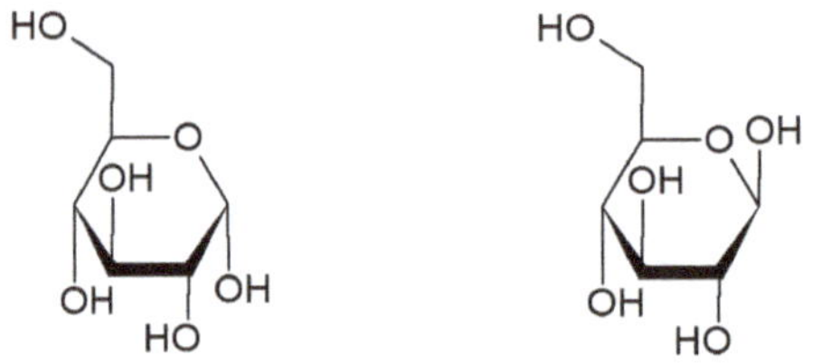

Figure 4.3: Starch and cellulose linkage of sugars.

The alpha linkage, as shown to the left of Figure 4.3, is exclusive to how starch binds one ring to the next. The beta link, shown on the right of Figure 4.3, is seen to be the dominant linkage that makes up cellulose. Although this seems like a small difference, it is one that is critical for organisms like yeast, which can consume only the alpha linked material, fermenting it into ethanol. Most cellulose that is converted to ethanol must first have some chemical pretreatment, such as an acid digestion, which breaks many of the ring-to-ring linkages. This begins the breakdown of the material but adds cost to the overall process.

Again, this may seem to be a small difference in an otherwise rather large molecule, certainly between two that are isomers, and that only differ at one carbon atom. Yet this difference is all that is necessary for one form of the carbo-hydrate, that from which cellulose is made, to be impossible for commercially available yeasts to degrade.

4.1.3 Nontraditional sources

A number of plant sources have been tried in the recent past, in a broad search for inexpensive, readily available plant sources from which ethanol can be produced. Those discussed below are not the only sources that have been tried in corporate or academic lab settings. But they are some of the various plants that have received attention because they have been found to have promising yields of bioethanol.

4.1.3.1 Algae
Recently, the use of algae to produce ethanol has been brought close to the commercial stage of production. Algae has been found to be a potentially good source for bio-diesel, because some varieties contain a significant amount of triglycerides that can be liberated and converted to diesel fuel. But the use of algae strains for bioethanol is

more recent. Depending on the type of algae and the growth conditions, though, claims have been made for production of several thousands of gallons of ethanol per acre, where corn usually produces 300–400 gallons per acre [22–23].

4.1.3.2 Switch grass

The use of switch grass – a weed plant that grows in the Midwestern United States – as a potential source for ethanol is promising enough that its use has been reported in the popular press [24]. Since it is not a food source, the use of it for ethanol does not put it in competition with any food. However, switch grass has never before been grown in fields and on a large enough scale to determine if the production of ethanol from it can make a positive difference in meeting the overall fuel needs of the world.

4.1.3.3 Agricultural residuals, fruit wastes – apple pressings

A large amount of research has gone into the use of residual plant materials from agriculture to produce bioethanol. Like switch grass, the main advantage of this is that these materials are already considered secondary to some main product, are not used for food for humans or animals, and thus often can be used without any conflicting interests. The disadvantage is the volume of them that can be generated, especially when compared to the amount of ethanol that is needed to meet overall fuel demands, again analogously like switch grass. Still, the use of waste agricultural materials holds promise for reducing the amount of other materials that are needed to produce the amount of fuel that will ultimately meet consumer demand.

4.1.3.4 Animal manure

Another form of agricultural waste that has been used to generate a biofuel is cow manure – and the fuel generated is methane. In farms which keep herds of milk cows in close proximity, their manure can be gathered and placed in a bio-digester, where methane that is generated is gathered for use. A bio-digester is generally a closed pit or receptacle with controlled entrances, and a controlled exit, the former where manure is put into the digester, the latter which gathers the methane. Figure 4.4 shows the basic design of a bio-digester.

In such a system, the methane is often called biogas and has been used to power the farm where it is generated. However, methane can also be a source of hydrogen gas, obtained through what is called hydrocarbon stripping. Hydrogen can then be used to hydrogenate ethylene to produce ethanol.

In the United States, if more power is generated than is used by a farm with bio-digesters, it is sold back to the local power company for some profit. Outlines and instructions for the production of a bio-digester can be found at the U.S. Environmental Protection Agency website [25]. Note that liquid and/or solid waste must still be removed at various times, as not all waste is converted to biogas.

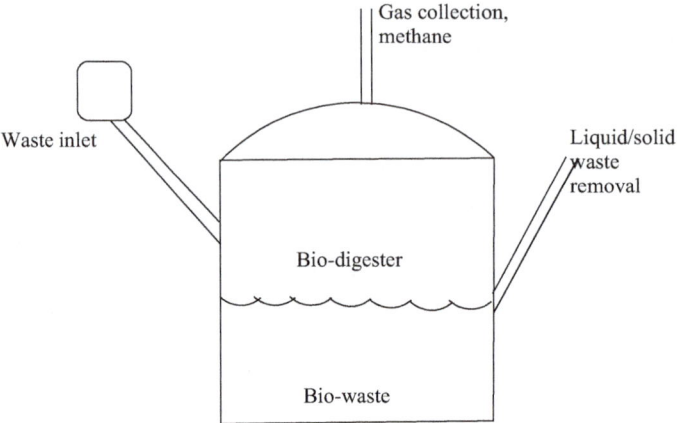

Figure 4.4: Bio-digester.

4.1.3.5 Bamboo

Like most plants, bamboo grows well in some parts of the world, but not in others. Where it does grow well, some types of bamboo can grow up to 24″ in a single day. The growth rate is such that significant amounts of bamboo, and what is called bamboo waste, are generated. Bamboo waste still does require treatment with a mildly alkaline solution to initiate breakdown of the cellulose, as the production process begins, but it can be a useful source for the production of bioethanol [26–28].

4.2 Biobutanol

Biobutanol has not been produced on as large a scale as bioethanol, nor has it been produced for as long, but it is produced in large part because it can be marketed as what is termed a "drop-in fuel." This means that existing internal combustion engines do not need to be adapted to the use of biobutanol, or to blends of it and traditional octane-based gasolines. It has been blended into traditional gasoline mixtures up to 12.5%, in terms of volume [29]. Figure 4.4 shows a flexible fuel designator that is seen on many automobiles today, although this can also be used when ethanol is the second fuel.

If biobutanol has a disadvantage, it is that its energy content is lower than traditional gasoline mixes, by up to 20%. Figure 4.6 shows the basic chemistry that produces butanol. Note that carbon dioxide is the coproduct.

The process is often called ABE, for acetone–butanol–ethanol, and the mix can be produced by *Clostridium acetobutylicum*, a microbial species. This traditionally has not produced a majority of butanol, and thus efforts continue to find or alter microorganisms to enhance the fraction of butanol that can be produced through

Figure 4.5: Flex fuel indicator.

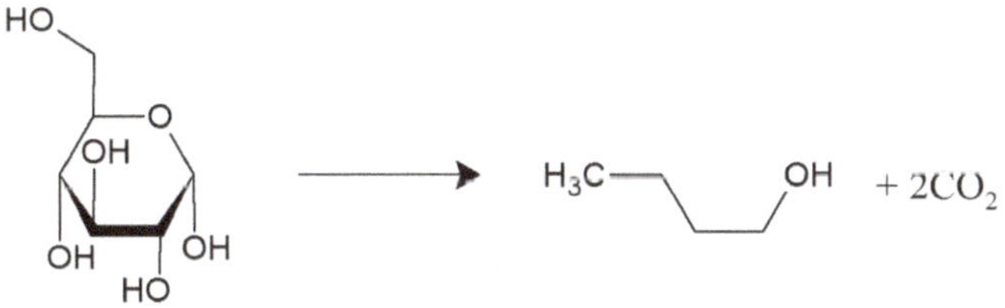

$$\text{(structure)} \longrightarrow H_3C-\text{---}\text{OH} + 2CO_2$$

Figure 4.6: Biobutanol production.

the ABE fermentation. As might be imagined, the separation of the butanol from other liquid products is a step that requires an input of energy, and thus adds to the cost of its production.

Perhaps the single greatest advantage of biobutanol when compared to bioethanol is the just-mentioned use as what is called a "drop-in fuel." This means that automobile manufacturers do not have to change their production lines to accommodate this fuel. This point may seem to be removed from the chemistry of producing biobutanol, but is in reality quite connected to it. Changing an automobile production line by creating a new type of engine requires a large input of time, energy, and personnel power.

4.3 Biodiesel

Diesel fuel is a mixture of isomers of what can be called middle-weight hydrocarbons. Diesel mixtures have been formulated especially for diesel engines, those which do not require both air and fuel in the combustion chamber – diesel requires only air in the chamber. Such engines are used in many trucks, larger construction vehicles, military vehicles, and some automobiles.

The isomeric mixtures that make up diesel fuel are not long enough molecules to be considered polymers but can be called waxes based on molecular composition. Saturated hydrocarbons make up roughly three quarters of diesel fuel mixtures, while a variety of aromatics make up the rest.

Because the composition of diesel fuel is very much like paraffins, diesel fuel is sometimes blended with small amounts of gasoline in cold climates because diesel by itself can wax and solidify. This varies depending on the region and the winter temperatures.

What has traditionally been called diesel fuel is now sometimes referred to as "petro-diesel" because its source is crude oil. When diesel fuel is made from renewable sources, which means plant or animal fats, it is called "biodiesel." This is now a large enough process that organizations are devoted to promoting it [30, 31].

We will discuss biodiesel here but will make a note that some methods of producing bio-diesel result in a pure enough product that for long-term storage approximately 1% of petro-diesel gets mixed into the fuel. This prevents it from degrading – rotting, really – because microorganisms can attack and consume the pure biodiesel.

4.3.1 Vegetable sources

While a wide variety of plant sources can be used to produce biodiesel, soybeans have seen the most use in the past few decades. Figure 4.7 shows the breakdown of triglycerides to 3 mol of fatty acid and 1 mol of glycerol. The fatty acids are then routinely converted to the methyl ester and used as biodiesel. The glycerol is not used in biodiesel. As a result, the increased production of biodiesel has kept the price of glycerol very low on world markets.

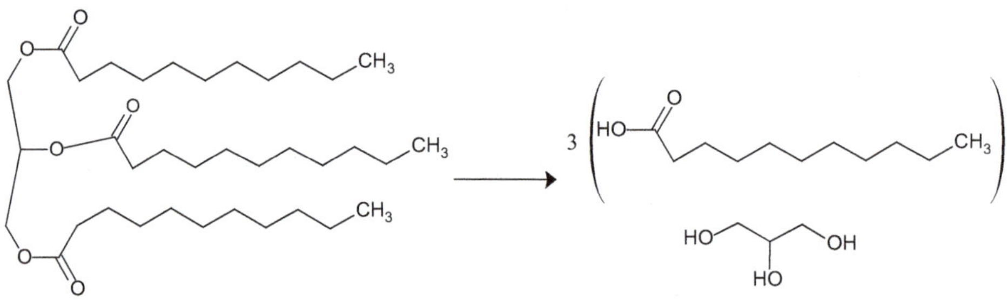

Figure 4.7: Fatty acid production.

Note that in Figure 4.7 we have shown each of the tails of the triglyceride as the same fatty moiety, thus becoming the same fatty acid. Depending on the plant or animal source, this does not have to be the case.

4.3.2 Animal sources

As well as plant sources such as soybeans, biodiesel can be made from a wide variety of animal fats. Excess animal fat is not routinely a food product, and thus the production of biodiesel does not compete with it for that function. Excess animal fat is used in several other industries, though, such as cosmetics and soaps [32]. The chemistry involves in producing biodiesel from animal fat is the same as that used with plant sources, as shown in Figure 4.7. However, the starting triglycerides in animal fats are not usually the same, as is mentioned in the comments about Figure 4.7. The end result is a form of biodiesel that has a wider variety of fatty esters in its mix than does that made from plant sources.

4.4 Future and challenges

Overall, the major challenges of biofuels are both producing enough of them to meet a significant portion of the existing demand for what are now petroleum-based fuels, and producing them without relying on food materials as their source. The former is a problem related to biology, plant growth, animal husbandry, and the availability of arable land. The latter is a question of social responsibility, since it invariably translates to taking food from roughly the poorest billion people in the world, and using it for transportation fuel for roughly the richest billion.

References

[1]   T. Standage. The History of the World in 6 Glasses, Bloomsbury Publishing, New York, 2005, ISBN: 978-0-80271-447-3.
[2]   Ethanol Producer Magazine. Website. (Accessed 6 July 2021, as: http://www.ethanolproducer.com/).
[3]   AzoCleantech. Website. (Accessed 12 April 2020, as: azocleantech.com).
[4]   American Soybean Association. Website. (Accessed 6 July 2021, as: https://soygrowers.com/).
[5]   Association of the Advancement of Industrial Crops. Website. (Accessed 6 July 2021, as: https://aaic.org/).
[6]   National Biodiesel Board. Website. (Accessed 17 July 2021, as: www.biodiesel.org).
[7]   National Corn Growers Association. Website. (Accessed 17 July 2021, as: www.ncga.com).
[8]   Pellet Fuels Institute. Website. (Accessed 17 July 2021, as: www.pelletheat.org).
[9]   Renewable Fuels Association. Website. (Accessed 17 July 2021, as: https://ethanolrfa.org).
[10]   Corn Refiners Association. Website. (Accessed 6 July 2021, as: https://corn.org/).
[11]   Growth Energy, America's Ethanol Supporters. Website. (Accessed 6 July 2021, as: https://growthenergy.org/).
[12]   Illinois Corn Growers Association. Website. (Accessed 6 July 2021, as: https://www.ilcorn.org/).
[13]   Illinois Soybean Growers Association. Website. (Accessed 6 July 2021, as: https://www.ilsoygrowers.com/).
[14]   Advanced Biofuels Canada. Website. (Accessed 6 July 2021, as: Advancedbiofuels.ca.).
[15]   EWABA | European Waste-to-Advanced Biofuels Association. Website. (Accessed, 6 July 2021, as: https://www.ewaba.eu).
[16]   Canadian Renewable Fuels Association. Website. (Accessed 6 July 2021, as: https://www.bio.org).
[17]   Bioenergy Australia. Website. (Accessed 6 July 2021, as: https://www.bioenergyaustralia.org.au).

[18] Bioenergy Association of New Zealand. Website. (Accessed 6 July 2021, as: https://www. bioenergy.org.nz).

[19] Southern African Bioenergy Association (SABA). Website. (Accessed 6 July 2021, as: http:// www.saba.za.org).

[20] The Brazilian Union of Biodiesel and Biojetfuel (Ubrabio). Website. (Accessed 6 July 2021, as: https://ubrabio.com.br).

[21] Mexican Network of Bioenergy (REMBIO) Website. (Accessed 6 July 2021, as: https://rembio. org.mx/en/).

[22] Algenol. Website. (Accessed 21 July 2021, as: algenol.com).

[23] C.E. De Farias Silva, A. Bertucco. Bioethanol from microalgae and cyanobacteria: A review and technical outlook. Process Biochemistry. 51(11), 2016, 1833–1842.

[24] Smithsonian Magazine. "The Next Generation of Biofuels Could Come From These 5 Crops", Website. (Accessed 21 July 2021, as: smithsonianmag.com 2017).

[25] Environmental Protection Agency. "Is Anaerobic Digestion Right For Your Farm?" Website. (Accessed 21 July 2021, as: epa.gov).

[26] J. Littlewood, L. Wang, R.J. Murphy. Techno-economic potential of bioethanol from bamboo in China. Biotechnology for Biofuels. 6(173), 2013 https://biotechnologyforbiofuels.biomedcentral. com/articles/10.1186/1754-6834-6-173.

[27] M. Kuttiraja, R.K. Sukumaran. Bioethanol production from bamboo (Dendrocalamus sp.) process waste. Biomass & Bioenergy. 59 Dec 2013, 142–150.

[28] H. Lu, X. Lin, B. He, L. Zhao. Enhanced separation of cellulose from bamboo with a combined process of steam explosion pretreatment and alkaline-oxidative cooking. Nordic Pulp and Paper Research Journal. 35(3), 2020, 386–399.

[29] U.S. Department of Energy, Alternative Fuels Data Center. Biobutanol. Website. (Accessed as: afdc.energy.gov).

[30] The National Biodiesel Board. Website. (Accessed 21 July 2021, as: https://www.biodiesel.org).

[31] Advanced Biofuels Canada. Website. (Accessed 21 July 2021,as: https://advancedbiofuels.ca).

[32] Farm Energy. Website. (Accessed 19 July 2021, as: https://farm-energy.extension.org).

Chapter 5
Nuclear power

5.1 Nuclear power – introduction

The production of electrical energy from radioactive fission has its origins in the production of atomic weapons during the Second World War. The Manhattan Project was the United States effort to produce an atomic weapon, although there were others at the time, in Great Britain, Canada, and both Germany and Japan – such as "Tube Alloys," the code name for the British project. At the conclusion of the war, not only had the scientific world come to the realization that the atom could be split (which had been hotly debated as recently as the 1930s), but a great deal had been learned about how isotopes of uranium could be separated and concentrated – the term generally used is "enriched" – and how this enriched uranium gave off heat. With the benefit of hindsight, it is not hard to realize that some lower level of enrichment could produce enough heat to generate steam and then to turn turbines, if only the reaction could be controlled. To be fair, however, this must have been an extremely large and impressive intellectual leap to the people who first made it.

Since that time, the development of nuclear power and nuclear power plants has become an important industry in many countries. There are now numerous governmental and nongovernmental organizations devoted to the promotion and understanding of nuclear power [1–25]. The United States currently has over 90 functional nuclear power plants, while France has 56, to give two examples. Since France is a smaller nation than the United States, just over 70% of the electrical power in France is provided by nuclear plants.

5.2 Nuclear fission

Radioactive fission is the splitting of an atom at its nuclear core into two smaller, lighter molecular weight atoms, often with the emission of some other small particles, such as neutrons. When this occurs, an enormous amount of energy, heat, is also given off.

Mathematically, the progression by which nuclear fission occurs is a matter of propagation of particles, usually in increments of three. What this means is:
1. A particle is hit with another particle of sufficient energy that the target particle is split.
2. The split particle results in two lighter elements, often krypton and barium (although there are also other possibilities), but three neutrons as well.
3. The neutrons have enough energy to each to further split an atom – now three atoms are being split.

https://doi.org/10.1515/9783110662276-005

4. Each of the three target atoms splits and again releases three further neutrons.
5. Now nine neutrons repeat the process.
6. Repeating this process over and *with no modification or control of neutrons* results in an atomic explosion when enough fissile material is concentrated and present in one place.
7. Repeating this process over and over with sufficient control of the flow of neutrons when the fissile material is in some liquid, such as water, results in enough heat that steam can be generated and made to turn turbines. This is how existing nuclear power is generated.

5.2.1 Reaction chemistry

Uranium exists in several different isotopes, with uranium-238 – often written as $^{238}_{92}U$ – being by far the most common one. Uranium-238 is referred to as being fissilely dead, however, and is not usable in power generation.

Uranium-235 is the fissile isotope of uranium. The predominant means by which the isotope uranium-235 undergoes fissile nuclear decay is shown in Figure 5.1. Note the three neutrons that are part of the decay products. As mentioned, these possess enough energy to repeat the reaction with three other $^{235}_{92}U$ atoms, thus promulgating the chain reaction.

$$^{235}_{92}U + {}^{1}_{0}n \rightarrow {}^{94}_{36}Kr + {}^{139}_{56}Ba + 3{}^{1}_{0}n$$ **Figure 5.1:** Uranium-235 decay.

It is noteworthy that there can be more than one fission pathway for uranium-235 to lighter elements. The reaction shown in Figure 5.1 is one that has been well studied. It is the moderation of the three neutrons which are given off that are of immediate concern. These neutrons have enough energy that if left unchecked in a nuclear reactor, the chain reaction can cause so much heat that what is called the core of the reactor could become hot enough to melt the nuclear fuel. This is the "melt down" that is brought up and mentioned in the popular press, as well as in many movies, in the event of any nuclear power-related accident.

5.2.2 From ore to fuel rods

Uranium is mined in various parts of the world, and in its natural state it is composed of more than one isotope. Uranium-238 is the predominant isotope and is referred to as fissilely dead. It is uranium-235 that must be separated from the ore and concentrated ultimately to produce nuclear power.

The original concentration of uranium-235 was done by gasifying samples of uranium as UF_6 and pushing the gas down long tubes. The molecules that reached the target first were higher in uranium-235 because this isotope is slightly lighter in overall mass than the predominant uranium-238. The basic reaction chemistry by which uranium-235 was gasified is shown in Figure 5.2.

$$U_3O_8 + HNO_3 \rightarrow UO_2(NO_3)_2$$

- This represents dissolving what is called "yellow cake" in nitric acid.
- Then solvent extraction follows to purify

$$UO_2(NO_3)_2 + NH_{3(solution)} \rightarrow (NH_4)_2U_2O_7$$

- This product is called ammonium diuranate

$$(NH_4)_2U_2O_7 + H_{2(g)} \rightarrow UO_2$$

- The oxide is treated with HF (hydrofluoric acid)

$$UO_2 + HF \rightarrow UF_4 + H_2O$$

- Finally, F_2 is added

$$UF_4 + F_2 \rightarrow UF_6$$

Figure 5.2: Production of UF_6.

The reason that two such dangerous materials – HF and elemental fluorine – were used to produce a particular gas is that fluorine is monoisotopic. Thus, the difference in mass between $^{235}_{92}U$ and $^{238}_{92}U$ versions of UF_6 is only that of the uranium, and never because of differences in fluorine isotopes.

More recently, the gas centrifuge method has been used to enrich uranium, since it is less energy intensive than the older gas diffusion process. High-speed centrifuges are utilized to separate the uranium isotopes. Multiple gas centrifuges are linked, in what is called a cascade, to produce successively higher concentrations of uranium-235.

Awareness of the centrifuge technique has been in the popular press in the relatively recent past because the use of this method by Iranian scientists may have been blocked by the introduction of a computer virus to the centrifuge computers. Speculation is that this was most likely affected by some element of the Israeli Defense Forces [26, 27].

The details of the production of what sometimes is still called "hex" – uranium hexafluoride (sometimes abbreviated DUF_6) – are not routinely written into a simplified reaction, such as in Figure 5.2. As mentioned, fluorine is used, despite its toxicity and reactivity, because it is monoisotopic. But also, and importantly, it is in the interest of governments that have the capability to perform either type of refining to guard the details of their operations jealously, so that other nations or rogue states and organizations do not acquire the capability to reproduce it, and thus to enrich uranium to higher levels, to what are called weapons grade.

5.2.3 Nuclear waste

5.2.3.1 DUF$_6$

Before discussing the treatment and storage of spent nuclear fuel, we should examine the production of significant amounts of depleted DUF$_6$ from the enrichment process. Currently, several hundred thousand tons of this waste products are contained and stored for an indefinite amount of time. Figure 5.3 shows the basic reaction chemistry if DUF$_6$ is exposed to moist or damp air over long periods of time.

$UF_6 + H_2O \rightarrow UO_2F_2 + HF$ **Figure 5.3:** Depleted uranium hexafluoride reactivity.

Both the products, uranyl fluoride and hydrogen fluoride, are toxic and can pose problems to their environment.

5.2.3.2 Fuel rods

The enriched uranium is reduced to metal and is used in a nuclear power plant in the form of fuel rods. These produce heat through radioactive decay, and are used to heat water to steam, which then turn turbines.

The location of each uranium fuel rod is carefully monitored, and its position within the core as well as its lifetime is tracked. When a rod is determined to be spent, it is removed from the core and monitored as a spent rod, essentially a waste product. Currently, in the United States, spent fuel rods are stored on or near reactor sites, where they are monitored indefinitely.

5.2.3.3 Control rods

The control rods of nuclear power plants are often made of elements such as boron or indium, as well as silver, cadmium, or hafnium. These are used because they have what is called a high-neutron cross-capture spectrum – they absorb neutrons well. Their working lifetime is also carefully tracked, so that plant personnel know when to change them out of the core, and when they are no longer useful for moderating and controlling power generation. At such time, each fuel rod must be safeguarded and monitored so that it does not release radiation to the surrounding environment. Currently in the United States, this also involves long-term storage of rods in water-filled cooling pools near the site of nuclear reactors.

The federal government proposal to move such waste to a storage facility under Yucca Mountain, Nevada, has been stopped. Concerns exist about the routes of travel to get spent nuclear fuel to the site, and the safety of populations along the way from nuclear plants to the Yucca Mountain site. As well, concerns exist about how a storage facility could be set up with safeguards that would last approximately 25,000 years, the estimated time for radioactive decay to levels that are not considered harmful.

5.2.4 Nuclear power plant operations

There are nearly 100 nuclear power plants functioning in the United States today, almost 60 functioning in France, and numerous others throughout the rest of the world. There is no single design for all plants. Indeed, what are called different generations of plants exist, with older plants being considered part of earlier generations, and with newer plants being built with greater numbers of overlapping safety devices and protocols.

The basic design of what can be called a pressurized water reactor is shown in Figure 5.4. It depends upon three loops of water, and none of which ever comingle with the other two. The basic steps of the plant are as follows:

1. Nuclear fuel, referred to as cores, are the center or heart of the plant. They give off radiation and heat.
2. The nuclear fuel is in water, which absorbs the heat.
3. When the water in the core loop boils, the steam can be used to turn turbines, which can then generate electricity.
4. The steam is recondensed to liquid water by the water in a second loop.
5. The water of the second loop is cooled by traveling through pipes that are immersed in a third loop of water. This third loop is connected to some large body

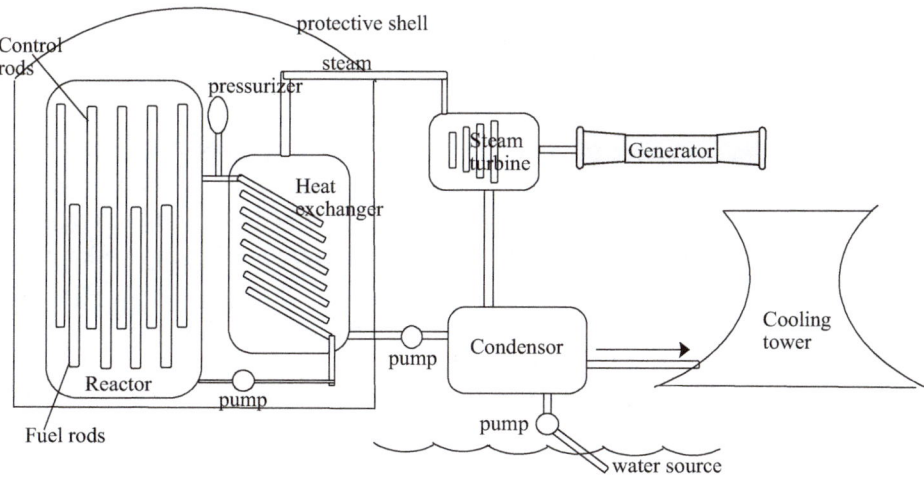

Figure 5.4: Basic scheme of a pressurized water nuclear power plant.

of water, such as a lake, running river, or ocean. This large body of water is always large enough to dissipate any residual heat from the second loop's water, into the greater aqueous environment.

Other designs have been utilized successfully for nuclear power plants but the pressurized water design is an established one that has been used many times over.

5.2.5 Nuclear power accidents

It may seem biased to devote a section of one chapter to the accidents that have occurred in an industry but to not do so for others. We do so here, however, because the nature of accidents at nuclear power generation facilities is such that they can produce radioactive pollution capable of lasting for centuries in some environment. As well, the general population has a more poignant fear of contamination from the results of an accident at a nuclear power plant than it does an accident at a coal-fired power plant or a wind turbine farm, simply because radiation can sicken or kill without anyone seeing, smelling, or in any way feeling its presence.

Arguably, the three most widely known nuclear power plant accidents are those of Three Mile Island in 1979, in Chernobyl in 1986, and at Fukushima in 2011. The first two were largely caused or compounded by human error, while the third was the result of an underwater earthquake and a resulting tsunami. In every case, including these three, but beyond them as well, improvements in design have been made in attempts to ensure that no further accidents occur.

5.3 Thorium-based power

It has been said that the reason the nuclear power of today is entirely based on uranium, as opposed to any other element, is because those working on the first atomic weapons realized that some uranium source enriched to a lower degree than that which was needed for a weapon could be used to generate heat – and thus generate steam and ultimately produce electric power. Professor Richard Feynman refers to this in an almost cavalier way in his first autobiography, "Surely You're Joking, Mr. Feynman" [28]. We might be venturing into the realm of speculation but it can also be said that the leaders of early atomic weapon programs, such as the Manhattan Project, might have had some foreshadowing of what their work would lead to, and thus strove for any peaceful use for enriched uranium as some form of atonement for the weapons they had created.

The result of determining how to isotopically enrich uranium is that all current nuclear power plants function by using it. However, there is another actinide element which has an extremely long half-life, and which can be used for power generation: thorium.

What can be called the end of a thorium-based nuclear power plant is very much the same as that for several other sources that generate energy. In some way, a turbine is turned, connected to a generator, and the system is then connected to existing power grids. One can also argue that the reactor design is not all that different from the basic design used in an established, uranium-based plant. Figure 5.4 shows a basic schematic for a uranium-235 nuclear power plant, and the following can serve as a compare-and-contrast list between the two.

1. The thorium-based core is the source of heat for a thorium plant.
2. Unstable uranium-233 is used to initiate the nuclear fission.
3. Control rods are still required to modify and control the heat emitted.
4. A fluid is heated – this does not have to be water, and has in several cases been a molten salt, such as a molten fluoride. Such a design is often called an MSR, for molten salt reactor, and one was run successfully for several years in the United States, at the Oak Ridge National Laboratory.
5. The heated liquid/molten salt is used to generate steam in a second loop.
6. The steam is pressurized enough to turn a turbine.
7. Turbines are connected to generators to produce electricity.
8. The steam that turned turbines is again condensed and recycled in its loop.
9. Water used to condense the steam is from a large natural body of water and is returned to that body.

Although we have mentioned the MSR design, there are several others. The following designs have been found to be feasible for electric power generation but are not currently in use:

1. Boiling light water reactor.
2. Fast neutron reactor.
3. Heavy water reactor.
4. High-temperature gas-cooled reactor.
5. Pressurized light water reactor.

The terms "light water" and "heavy water" refer to the hydrogen atoms in the water. Light water is common H_2O, and heavy water has deuterons in place of protons, making it D_2O.

$$^{232}_{90}Th + ^{1}_{0}n \rightarrow ^{233}_{90}Th \rightarrow ^{0}_{-1}e + ^{233}_{91}Pa \rightarrow ^{233}_{92}U + ^{0}_{-1}e$$

Figure 5.5: Production of $^{233}_{92}U$.

The chemical reaction that illustrates the production of uranium-233 is shown in Figure 5.5. Note that the thorium fuel must capture a neutron to begin the transmutation process that leads to the fissile uranium-233.

Thorium is usually obtained from monazite ores, a phosphate mineral of varying composition, which has in the past been mined for their lanthanide components (aka, their rare earth elements, or REEs). Different ore sources of monazite have different amounts of thorium in them. The United States currently imports thorium from both India and France [29].

The means by which thorium is extracted from monazite is relatively complex simply because, as mentioned, monazite is always composed of several REEs and thorium as the cations. An example formula for a monazite ore batch might be: (La,Ce,Nd,Th) (SiO_4,PO_4). The refining of rare earths has traditionally been the economic driving force for extraction of elements from monazite, since elements like cerium have established uses (in alloys, for instance), as does neodymium (in strong, permanent magnets). The United States Geological Survey Mineral Commodity Summaries 2021 notes: "Globally, monazite was produced primarily for its rare-earth-element content, and only a small fraction of the byproduct thorium produced was consumed. India was the leading producer of monazite. Thorium consumption worldwide is relatively small compared with that of most other mineral commodities" [29]. Figure 5.6 illustrates a simplified chemistry for thorium isolation from some lanthanide-containing ore.

$$LnO_{x(s)} + H_2SO_{4(aq)} \rightarrow LnSO_{4(aq)}$$

$$LnSO_{4(aq)} + NaOH_{(aq)} \rightarrow Th(OH)_{4(s)} + LnSO_{4(aq)}$$

Figure 5.6: Thorium isolation.

Ln is used to represent the mix of REEs and thorium in the first equation. Further, the reaction requires heating to result in a solution of lanthanide – thorium sulfates. Importantly, following this, the reaction conditions require adjustment to an acidic pH range of 3–4. This allows the thorium hydroxide product to precipitate from solution, which still contains REEs.

5.4 Nuclear fusion

The promise of extracting energy in a controlled manner from some form of nuclear fusion is an amazing one, the promise of a virtually limitless source of energy. First conceived of in the 1950s, significant work was put into the idea up to the 1970s, and does continue today. Unfortunately, the harnessing of nuclear fusion in some controlled manner still has significant hurdles to overcome, including economic ones. A cynical comment about the process has been: "Fusion power is about forty years away, and forty years ago was about forty years away." Therefore, any sort of commercial use still appears to be in the future.

The process of fusion involves combining two nuclei, generally small ones, and through the synthesis of a new, heavier element capturing the energy created. Figure 5.7 shows the simplified reaction chemistry for the fusion of deuterium and tritium, the two heavy isotopes of hydrogen that have proven promising as isotopes which can be fused.

$$_1^2D + _1^3T \rightarrow _2^4He + _0^1n$$

Figure 5.7: Deuterium–tritium fusion.

The total mass of the resulting helium atom is less than the combined mass of the two starting atoms. Einstein's well known equation $E = mc^2$ can in this situation be more completely written $E = \Delta mc^2$, where the delta is the change in mass and the energy given off is related to this change.

Currently, a major challenge to any sustained nuclear fusion as a source of energy is the energy required to force two particles, such as deuterium and tritium atoms, to fuse. What is called the Coulomb repulsion between the nuclei is extremely large and must be exceeded if the two are to fuse. Nature does this in the form of stars such as our sun, which have enormous gravity pushing nuclei together. Doing this on the Earth, in a controlled manner, has proven to be a goal that has not yet been achieved in a sustainable manner.

5.4.1 Tokamak

Advances toward a fusion reactor have taken several forms. One prominent example is the tokamak, which is a toroidal-shaped device that confines a plasma in the shape of a torus using a strong magnetic field. This magnetic field is the means by which an extremely hot plasma is confined and controlled, in which the fusion of small isotopes like deuterium and tritium occurs. Currently, no plant design, whether a tokamak or other, has been able to produce more energy than has been put into it. The International Thermonuclear Experimental Reactor has become the largest effort directed to fusion power [30].

References

[1]   World Nuclear Association (WNA). Website. (Accessed 21 July 2021, as: http://www.world-nuclear.org/).
[2]   European Nuclear Society (ENS). Website. (Accessed 21 July 2021, as: http://www.euronuclear.org/).

[3]   Canadian Nuclear Association (CNA). Website. (Accessed 21 July 2021, as: https://cna.ca/).

[4]   Institute of Nuclear Power Operations (INPO). Website. (Accessed 21 July 2021, as: http://www.inpo.info/).

[5]   Institute of Electrical and Electronics Engineers (IEEE) Nuclear and Plasma Sciences Society (NPSS). Website. (Accessed 21 July 2021, as: http://ieee-npss.org/).

[6]   American Nuclear Society (ANS). Website. (Accessed 21 July 2021, as: http://www.ans.org/).

[7]   Associação Brasileira para o Desenvolvimento de Atividades Nucleares (ABDAN). Website. (Accessed 21 July 2021, as: http://abdan.org.br/).

[8]   China Nuclear Energy Association (CNEA). Website. (Accessed 21 July 2021, as: http://www.china-nea.cn/).

[9]   Chinese Nuclear Society. Website. (Accessed 21 July 2021, as: http://www.ns.org.cn/).

[10] World Association of Nuclear Operators (WANO). Website. (Accessed 21 July 2021, as: http://www.wano.info/en-gb).

[11] Women in Nuclear (WIN). Website. (Accessed 21 July 2021, as: http://www.win-global.org/).

[12] Nuclear Industry Association of South Africa (NIASA). Website. (Accessed 21 July 2021, as: http://www.niasa.co.za/).

[13] Minerals Council of Australia – Uranium Forum. Website. (Accessed 21 July 2021, as: http://www.minerals.org.au/resources/uranium).

[14] Korea Nuclear Energy Agency (KNEA). Website. (Accessed 21 July 2021, as: http://www.keia.or.kr//).

[15] Joint Institute for Nuclear Research (JINR). Website. (Accessed 21 July 2021, as: http://www.jinr.ru/main-en/).

[16] Nuclear Energy Institute (NEI). Website. (Accessed 21 July 2021, as: https://www.nei.org/).

[17] Gateway for Accelerated Innovation in Nuclear (GAIN). Website. (Accessed 21 July 2021, as: https://gain.inl.gov/SitePages/Home.aspx).

[18] Argonne National Laboratory. Website. (Accessed 21 July 2021, as: http://www.anl.gov/).

[19] Idaho National Laboratory. Website. (Accessed 21 July 2021, as: https://www.inl.gov/).

[20] Oak Ridge National Laboratory. Website. (Accessed 21 July 2021, as: https://www.ornl.gov/).

[21] Southern Nuclear. Website. (Accessed 21 July 2021, as: https://www.southerncompany.com/our-companies/southern-nuclear.html).

[22] Generation Atomic. Website. (Accessed 21 July 2021, as: http://www.generationatomic.org/).

[23] Terrestrial Energy. Website. (Accessed 21 July 2021, as: https://terrestrialenergy.com/).

[24] NAC International. Website. (Accessed 21 July 2021, as: http://www.nacintl.com/).

[25] American Public Power Association. Website. (Accessed 21 July 2021, as: https://www.publicpower.org/).

[26] D. Kushner. The Real Story of Stuxnet. IEEE Spectrum. 26 Feb 2013 Website Accessed 22 July 2021 as http://spectrum.ieee.org/telecom/security/the-real-story-of-stuxnet.

[27] An Unprecedented Look at Stuxnet, the World's First Digital Weapon, Wired, 3 November 2014. Website. (Accessed 22 July 2021, as: wired.com/2014/11/countdown-to-zero-day-stuxnet).

[28] R. Feynman. Surely You're Joking, Mr. Feynman, W.W. Norton, 1985, ISBN: 0-393-01921-7.

[29] U.S. Geological Survey. Mineral Commodity Summaries 2021, downloadable.

[30] International Thermonuclear Experimental Reactor, ITER. Website. (Accessed 23 July 2021, as: iter.org).

Chapter 6
Hydropower

6.1 Dams – introduction

The taming of streams and rivers by damming them, in order to derive some useful work from the movement of the water, is another technique that goes back to ancient times. If we include in this idea the construction of relatively small dams and water wheels, using what today would be considered very primitive tools, there is evidence that dam construction is actually millennia old [1]. The oldest dams in the world appear to go back over 2000 years. While several might claim to be the oldest, using just one example, the Kallanai Dam in Tamil Nadu, India, was built in the second century for the control and diversion of water, so that fields could be irrigated and so that seasonal drought and flooding would be minimized. Perhaps obviously, the use of dams to create electricity is a much more recent phenomenon, having started in the late nineteenth century. Specifically what is believed to be the oldest hydroelectric power plant began operations in 1882 in Appleton, Wisconsin. This is the Vulcan Street Plant. Since that time, dams as sources of hydroelectric power have become so prevalent that numerous organizations exist to promote their use, and awareness of their benefits to the general public [2–19].

The oldest waterwheels also go back centuries, and possibly millennia. What is believed to be one of the oldest tidal waterwheels is one that was archaeologically excavated in Northern Ireland, and is called the Nendrum Monastery Mill. It appears to have run not on the flow of river water, but on tidal power. But what appears to have been the oldest of all waterwheels seems to have been built in ancient Mesopotamia, over 2000 years ago.

6.1.1 Dam requirements

When considering the construction of a major modern dam, the obvious starting point is the river that will be channeled. As well, the kinetic flow of water through an existing channel in the river, and through some reduced channel in which turbines will be placed, is critical. From the point of view of the materials chemistry that is required, however, the amount of cement and concrete is one of the major factors. The turbines, generators, and other equipment required to convert the flow of water into electrical energy is another. Figure 6.1 shows a simple schematic for a dam and hydroelectric power plant.

https://doi.org/10.1515/9783110662276-006

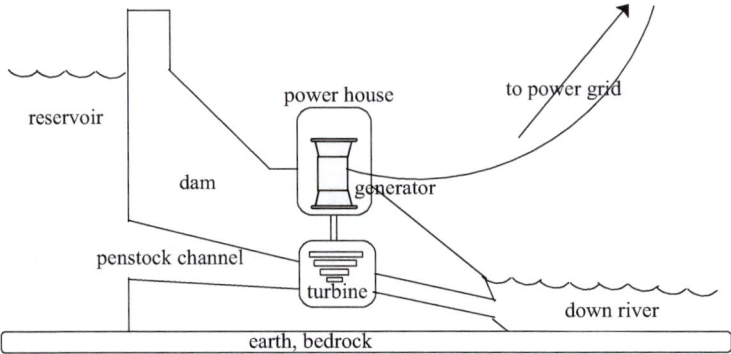

Figure 6.1: Dam schematic.

One aspect of all dams that may seem trivial or obvious, but that is also critical, is that the width of the dam at the lower level, or base, where the turbines are located, must be significantly thicker than that at the top, as indicated in Figure 6.1. This is so that the dam can withstand the constant pressure of the built-up water behind it, and so that flow through the sluice or penstock can be effectively regulated to turn the turbines without mishap.

Since the pouring of concrete for a dam can take from months to years, an entire infrastructure has to be built at or near the work site, so that workers can live near the site, trucks or railcars of equipment or material can be brought to the site, and all the equipment required to pour, shape, and form the dam can be brought to the correct place.

Figure 6.2 shows a portion of the Hoover Dam, spanning the Nevada–Arizona state borders, and the man-made Lake Meade behind it, giving both some idea of the scale of the construction and an indication that it is more than simply an enormous plug or diversion of a river. The individuals seen near the bridge give an idea of the size of the dam.

6.1.2 Dam locations worldwide

The use of hydropower has spread quickly and widely in the twentieth century; and dams are now found on all six inhabited continents. In the United States, much of the general public is aware that Hoover Dam, on the Colorado River, was the largest in the world at the time it was built at 726.4 feet. But in 1968, it became the second largest, since the Orville Dam in California is 770 feet. As well, the Three Gorges Dam in China is currently the world's largest. No current projects exist that threaten to challenge this.

A listing of the biggest dams in the world is shown in Table 6.1, along with the dam near Niagara Falls, on the US side of the Niagara River. This latter is listed here because

Figure 6.2: View at Hoover Dam.

the original construction is one of the oldest hydroelectric power generating stations in the world, and remains a useful point of comparison. Note the power listed for each.

Table 6.1: Largest hydroelectric power dams in the world.

Name	Location	River	Capacity, installed (MW)
Three Gorges	Hubei, China	Yangtze	22,500
Itaipu Dam	Brazil and Paraguay	Parana	14,000
Xiluodu	China	Jinsha	13,860
Guri	Venezuela	Caroni	10,235
Tucurui	Brazil	Tocantins	8,370
Belo Monte	Brazil	Xingu	8,176
Grand Coulee	USA	Columbia	6,809
Xiangjiaba	China	Jinsha	6,448
Longtan Dam	China	Hongshui	6,426
Sayano-Shushenskaya	Russia	Yenisei	6,000
Robert Moses Niagara Hydroelectric Power Station	USA	Niagara	2,625

The power listed for each dam can fluctuate based on the seasonal fluctuation of the river where the dam is located. Also, it is perhaps obvious that a river can be dammed more than once. The Colorado River, as mentioned the site of the Hoover Dam, has 15 dams along its main course and over 100 along the various tributary rivers that feed into it.

Additionally, dams can be multipurpose in nature, meaning they can be used for the generation of electricity, and at the same time used for the storage of water [20]. This is certainly important in areas that are generally dry, or that undergo droughts. Again, the Colorado River system serves as a source of water for human use, but so do many other dams. Those built by and planned by the government of Turkey serve as further examples [21].

6.2 Run-of-river hydropower

When a full dam, inclusive of a reservoir lake, is not needed to produce electric power, the river upon which a hydroelectric power plant is built must be fast enough flowing that any turbines in the river can run without the pent up potential energy of water elevated above a turbine. This type of hydroelectric power is often referred to as run-of-the-river, sometimes abbreviated ROR. The design of each plant can differ, but the common feature is that they have no reservoir, or that they have what is called pondage, a small water storage facility. Beyond this, the flow of water turns turbines, which are connected to electric generators, functioning exactly as dams with reservoirs function.

The western United States and western Canada have terrain well-suited to the construction of ROR power stations. British Columbia has numerous such dams and appears poised to expand the existing number. In the United States, the Chief Joseph Dam on the Columbia River in Washington State is a large ROR that has been in operation for over four decades. It must move the water that comes downriver from the Grand Coulee Dam at approximately the same rate of flow that it receives it. The next dam downriver is the Wells Dam. If water flows to the dam in excess of the capability of its 27 generators, the gates to its spillways are opened so that the river can continue to flow at a natural rate, having little capacity to store water.

A major advantage of ROR hydroelectric power is that it can have a minimal effect on the river, certainly when compared to a traditional dam. This is especially important when endangered fish and wildlife depend upon the river. The major disadvantage for this type of power generation is that ROR projects are not considered as reliable as a dammed river. They are subject to seasonal fluctuations in the river's water flow, based on rainfall and snow pack melt.

6.3 Tidal power

The Earth and the Moon have been in what has rather romantically been called a celestial dance for hundreds of millions, perhaps billions, of years. Because of the Moon's pull on the Earth, tides constantly move in and out, generating an extremely large amount of energy as they do so. Virtually all of it is unused by humanity. The idea of harnessing the power of tidal movements is new, but has gained significant ground in the past few decades. There are currently several tidal power stations in use throughout the world, utilizing a number of different designs. Since the technology for this form of hydroelectric power production is still new, there is no one design that is agreed upon as the best, although all depend simply upon the kinetic energy of water as the tides comes in and goes out. Table 6.2 provides some examples of existing tidal power stations.

Table 6.2: Areas of tidal power stations.

Name	Location	Capacity	Website	Comment
Annapolis Royal Generating Station	Bay of Funday, Nova Scotia	20 MW	https://www.nspower.ca	Since 1984
Jiangxia Tidal Power Station	Hangzhou, China	3.2 MW		Since 1985
Jindo Udolmok	South Korea	90 MW		Since 2009
Kislaya Guba	Barents Sea, Russia	1.2 MW		
Race Rocks	Vancouver, Canada		racerocks.ca	Shut down
Rance	LaRance, France	240 MW	https://www.edf.fr	Started in 1960
SeaGen	Strangford Lough, N. Ireland	1.2 MW		Since 2008
Sihwa	South Korea	254 MW	http://energy.korea.com/archives/6887	Since 2011

It is rather easy to become confused as to the meaning of so many large numbers, in terms of megawatts, and what they mean. As a point of reference, the Sihwa station in South Korea is large enough that its annual output is large enough to power the requirements of the residences of a 500,000 person city. This gives both an idea of the output of the station, and also a realization of how much more tidal power must be harnessed if humanity expects to derive any significant percentage of its power from it, for cities, for residences, for corporate operations, for the myriad other power requirements now before us.

6.3.1 Tidal power designs

Such stationary facilities for the capture and use of tidal power are designed based on the area in which they will be constructed, or in which they will be situated. Some tidal escarpments are basically a form of dam, which traps water at a certain time of day with what is called the barrage, then spills it back at some later time. A basic schematic of this is shown in Figure 6.3.

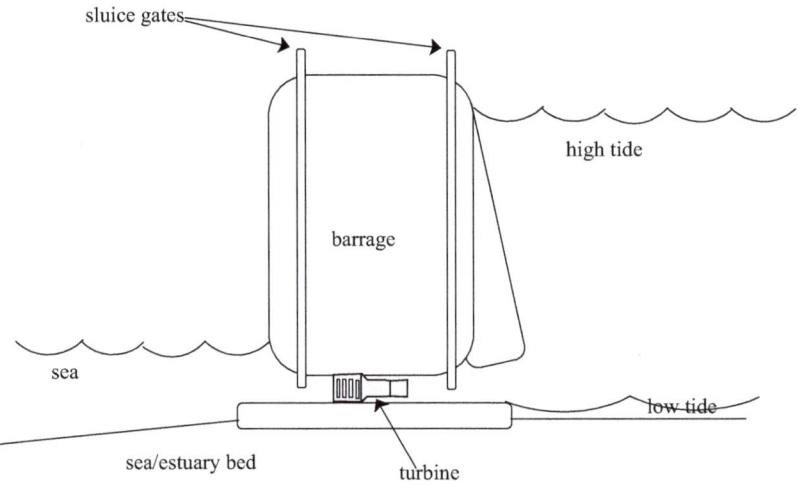

Figure 6.3: Schematic of tidal power generation station.

As can be seen, the chemistry involved in the construction of such a project is a combination of the cement and solid materials that will be needed, as well as the materials required for the turbines, sluice gates, and other moving parts that will be installed, as well as the materials that are required to ensure the concrete barrage does not degrade over the course of time. Connecting to an existing power grid, or to an energy storage station (a battery hub), is a traditional operation of connecting a turbine to some form of generator. This is very much like a hydroelectric dam that generates electricity.

6.3.1.1 Tidal turbine
What are called tidal turbine power plants are, as the name implies, turbines in an estuary that basically stand alone, and are not part of some greater barrage and

damming project. Electric power is produced by the turning of turbine blades as water flows in or out during the tidal cycle. Once again, several different designs can be utilized, as long as the end result is that the design produces electricity. Figure 6.4 shows a basic schematic.

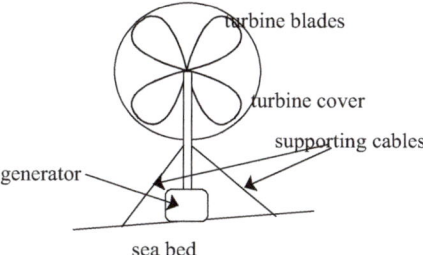

Figure 6.4: Schematic of a tidal turbine.

Note that Figure 6.4 shows a bottom-mounted turbine. Turbines can also be anchored or tethered by cables, and thus have more freedom of movement. As well, a system can be floating, as discussed below.

Another means of capture of the energy of tidal power, a floating tidal energy system, looks and functions much more like a ship than a dam. Advantages of such an operation are that the station is generally smaller, more mobile, and less intensive as far as materials are required to construct it. The disadvantage to all turbine systems is that their power output is nowhere near as high as a large, stationary tidal dam, meaning more units are required to equal the same energy output.

An example of such a floating tidal power generating station design is shown in Figure 6.5. It should be noted that this example is just that an example. Floating tidal energy systems are novel enough that there is not yet any standard design. Should they prove to be economically profitable, they will undoubtedly continue to evolve in the coming years, with the end result being the capture of greater amounts of energy. Note that even though a tether cable is not shown, such floating power stations need to be tethered to the sea bed or some other stationary object, so they have freedom of movement. The "catamaran-like" design ensures maximum stability in the littoral waters, and the turbines can be submerged or elevated out of the water, if needed [22].

The first of these floating tidal platforms have actually been built decades ago, but none have provided continued electricity production. The cost of construction and maintenance has in the past not been competitive with established means of power production, such as coal. Those in trial stages today have been made much more affordably, and electricity production from them, expected to be years or decades long, is just beginning.

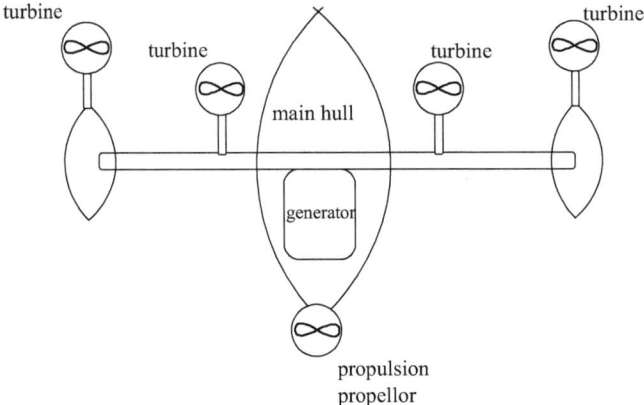

Figure 6.5: Floating tidal power generating station, top view.

6.3.2 Areas of large tidal difference

The littoral areas that show the most potential for tidal power are those with large differences in high and low tides, or those with fast running water during the change in the tide. These areas have the greatest potential for the height difference of incoming and outgoing water to be harnessed and used for energy. Table 6.3 is a list of some of the highest tidal areas in the northern hemisphere [23].

Table 6.3: Areas of large tidal differences.

Name	Tidal height (ft)	Comments
Bay of Fundy, Burntcoat Head, Nova Scotia	38.4	
Horton Bluff, Avon River, Minas Basin, Bay of Fundy, Nova Scotia	38.1	Bay of Fundy, considered one of the seven wonders of the Northern Hemisphere
Amherst Point, Cumberland Basin, Bay of Fundy, Nova Scotia	35.6	
Parrsboro – Partridge Island – Minas Basin, Bay of Fundy, Nova Scotia	34.4	
Hopewell Cape, Petitcodiac River, Bay of Fundy, New Brunswick	33.2	
Leaf Lake, Ungava Bay, Quebec	32.0	

Table 6.3 (continued)

Name	Tidal height (ft)	Comments
Port of Bristol – Avonmouth – Britain	31.5	Also site of the Avonmouth Port wind farm
Grindstone Island, Petitcodiac River, Bay of Fundy, New Brunswick	31.1	
Spencer Island, Bay of Fundy, Nova Scotia	30.5	
Newport, Bristol Channel, Britain	30.3	
Sunrise, Turnagain Arm, Alaska	30.3	Location of several tide stations
Anchorage, Alaska	26.2	Pilot projects in Cook Inlet
Cobscook Bay, Maine	20.0	Home of Reversing Falls Park

According to NOAA:

> Increased tidal ranges in these areas are created by the positions and configurations of the continents in the northern hemisphere. In the higher latitudes of the northern hemisphere, the continents of North America, Europe, and Asia are pressed closer together. This "constriction" of the oceans creates the effect of a higher range of tides. [23]

At this early stage in the development of tidal power, the height of tidal flows is perhaps obviously important. But as these technologies develop and mature, areas with smaller tidal differences in height may become viable for producing energy [23, 24].

6.4 Wave power

The term "wave power" may seem very close to synonymous with that of tidal power, but the two phenomena are different. Wave power is independent of tidal currents and is a means of producing electricity from wind-generated waves. Much like tidal power operation, however, wave power is in its earliest stages at the moment.

The first wave generator to reach commercial viability is called the Islay LIMPET – short for Land Installed Marine Power Energy Transmitter – which was connected to Britain's power grid in 2000, although it was decommissioned in 2012. It functioned as follows:

1. An inclined concrete hollow tube, with one end positioned underwater, allows water in when a wave surges.
2. The incoming water compresses air trapped in one end of the tube.

3. When the wave, and the water in the tube, is outgoing, the air expands in the tube.
4. This compression and expansion of air occurs in the presence of turbines, which can rotate in either direction.
5. The turbines are connected to generators (as with many of the energy-producing devices we have seen), which then produce electricity.

Importantly, this is only one design by which energy can be produced through the movement of waves. All such devices which do so can be called wave energy converters (WEC). Several designs have been tried, although proof of concept does not always translate into a commercial WEC. Recently, the Perth Wave Energy Project has been able to produce energy and to use it to desalinate seawater, utilizing a tether to force a pump anchored on the seafloor to compress and expand with the waves. The entire apparatus is designed to stay entirely submerged [25, 26].

6.5 Thermocline, ocean thermal energy conversion

The ocean exists in layers of different temperature and different aqueous salinity, and energy can be extracted with the proper placement of turbines utilizing the temperature difference of two different layers. This idea actually has a fairly long history but has not yet become a major contributing source of energy for the developed nations of the world. For example, the Hawaiian Islands have the potential for numerous power plants to be established that use what is generally called ocean thermal energy conversion (OTEC) as a source of power, as do the Japanese home islands, and Great Britain and Ireland. The costs of constructing such power plants, however, have thus far not been competitive with coal and nuclear power.

6.5.1 OTEC plant design

The design of a plant to harness energy from the temperature difference of different layers of ocean water always requires some piping system to bring deep, cold water to the surface. Additionally, warm surface water must be piped into the plant. Also, water must eventually be returned to the ocean.

In addition to the heat transfer that will produce energy, OTEC plants can be designed to produce desalinated water.

6.5.2 Chemical requirements, heat transfer

Cool water coming up from a deep part of the ocean (at approximately 4 °C) and warm water coming from the ocean's surface (at approximately 25–27 °C) do not automatically produce energy simply upon mixing. Figure 6.6 shows a schematic of how energy is produced, and how a second liquid is involved – a low-boiling liquid such as a refrigerant, or possibly ammonia. In sequential steps, it can be described as follows:

1. Cold ocean water (4 °C) is pumped from deep depths, ~1 km.
2. A low boiling point fluid is housed in a container or loop which has maximum exposure to the cold water intake, and makes contact as a gas.
3. The low boiling fluid is pumped to a portion of the loop where it has maximum exposure to warm water (25–27 °C).
4. Warm surface water is pumped into the OTEC chamber, where it undergoes maximum exposure to the low boiling liquid, vaporizing it.
5. The vaporized fluid is pushed through a turbine, then recooled.
6. The turbine is connected to a generator.
7. Cables from the generator to a shore station transmit power to land.

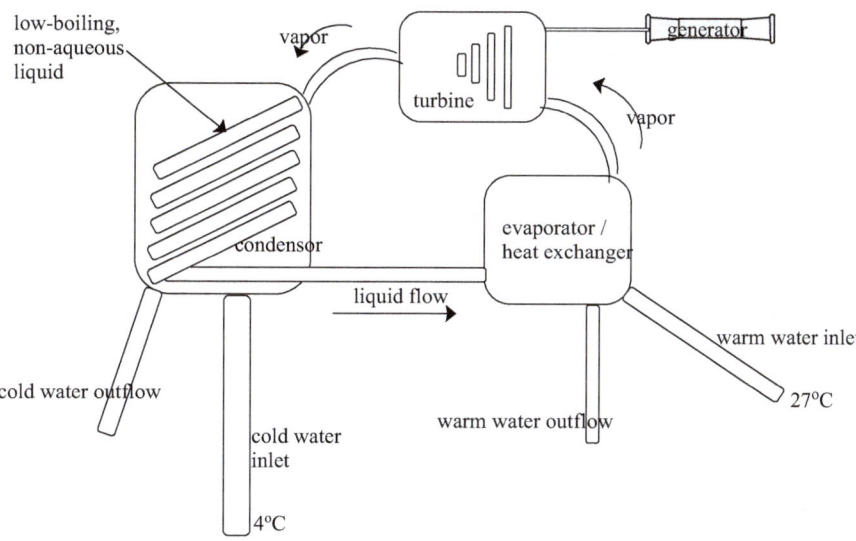

Figure 6.6: OTEC plant design.

6.5.3 Construction and maintenance challenges

OTEC plants have been constructed that have produced power, and from these a great deal has been learned about the challenges of continued operation and maintenance. The most obvious is probably the survivability of the plant during times of intense storm or any sort of bad weather. As well, marine organisms can build up on nonmoving parts over the course of time, fouling parts and limiting the usefulness of the operation. All this is important to learn as moves are made to some sort of full-time OTEC plant or plants that provide power continuously [27, 28].

References

[1]   The world's oldest dams still in use. 20 October 2013. Website. (Accessed 24 July 2021, as: www.water-technology.net/features/feature-the-worlds-oldest-dams-still-in-use).

[2]   International Hydropower Association. Website. (Accessed 21 July 2021, as: https://www.hy dropower.org).

[3]   National Hydropower Association. Website. (Accessed 23 July 2021, as: https://www.hydro. org/about/).

[4]   World Commission on Dams. Website. (Accessed 23 July 2021, as: https://www.international rivers.org/campaigns/the-world-commission-on-dams).

[5]   European Small Hydropower Association. Website. (Accessed 23 July 2021, as: https://ec.eu ropa.eu/energy/intelligent/projects/en/partners/european-small-hydropower-association-0).

[6]   Hydroworld. Website. (Accessed 23 July 2021, as: http://www.hydroworld.com/index.html).

[7]   Renewable Energy World. Website. (Accessed 23 July 2021, as: http://www.renewableenergy world.com/hydropower/top-news.html).

[8]   International Hydropower Association. Website. (Accessed 23 July 2021, as: https://www.hy dropower.org/congress/home).

[9]   International Center on Small Hydro Power. Website. (Accessed 23 July 2021, as: http://www. inshp.org/default.asp).

[10]   Indian National Hydropower Association. Website. (Accessed 23 July 2021, as: http://www. cbip.org).

[11]   National Hydropower Association. Website. (Accessed 23 July 2021, as: https://www.hydro. org/).

[12]   International Hydropower Association. Website. (Accessed 23 July 2021, as: https://www.hy dropower.org/).

[13]   Smart Hydropower. Website. (Accessed 23 July 2021, as: https://www.crunchbase.com/orga nization/smart-hydro-power).

[14]   Hydro Quebec. Website. (Accessed 23 July 2021, as: http://www.hydroquebec.com/about/ mission-activities.html).

[15]   Canyon Hydro. Website. (Accessed 23 July 2021, as: http://www.canyonhydro.com/resour ces.html).

[16]   British Hydropower Association. Website. (Accessed 23 July 2021, as: http://www.british-hydro.org/).

[17]   BC Hydro Power Smart. Website. (Accessed 23 July 2021, as: https://www.bchydro.com/ index.html).

[18]   Manitoba Hydro. Website. (Accessed 23 July 2021, as: https://www.hydro.mb.ca/).

[19]   AD Hydro Power Limited. Website. (Accessed 23 July 2021, as: http://adhydropower.com/).

[20]   Hoover Dam. Website (accessed 4 November 2019, as: https://www.usbr.gov/lc/hooverdam/).

[21]   Ataturk Dam, Anatolia, Turkey. Website. (Accessed 23 July 2021, as: www.water-technology. net/projects/ataturk-dam-anatolia-turkey).

[22]   Smithsonian, April 2020, pp.45–52.

[23]   Ocean Flow Energy, Ltd. Website. (Accessed 25 July 2021, as: www.oceanflowenergy.com).

[24] National Oceanic and Atmospheric Administration (NOAA), National Ocean Service (NOS). Website. (Accessed 25 July 2021, as: https://oceanservice.noaa.gov).

[25] Perth Wave Energy Project – Australian Renewable Energy. Website. (Accessed 25 July 2021, as: https://arena.gov.au).

[26] Carnegie Clean Energy. CETO Technology. Website. (Accessed 25 July 2021, as: https://www.carnegiece.com).

[27] U.S. Energy Information Agency. Hydropower explained: Ocean thermal energy conversion. Website. (Accessed 25 July 2021, as: https://www.eia.gov/).

[28] Makai Ocean Engineering. Website. (Accessed 25 July 2021, as: https://www.makai.com/ocean-thermal-energy-conversion/).

Chapter 7
Geothermal energy

7.1 Geothermal energy – introduction

The surface of the Earth has been described by some to be large plates of crustal material slowly floating on a sea of molten rock. The energy produced by this movement can be seen when volcanoes and geysers breach the crust and vent to the atmosphere. Throughout history only a minuscule amount of this massive reserve of energy has ever been captured. However, in the relatively recent past – starting in 1958 in New Zealand at the Wairakei Power Station – geothermal energy has been harnessed either through direct capture of the energy in material that escapes to the Earth's surface, or through what are called injection wells, to deep areas of high heat.

In all cases, the chemistry involved in geothermal energy is that of heat transfer. This transfer is routinely from a very hot subterranean spot to some cooler point at or near the surface. During this transfer, some movement of a gas or liquid means electricity can be generated through turning turbines, or in other ways capturing the energy of the hot portion of the system.

While geothermal energy remains a relatively small part of the overall energy profile used by nations to provide electricity to their populations, its use continues to grow; and there are now professional organizations devoted to its use and awareness [1–9].

7.2 Geographic locations

Geothermal energy is often associated with the edges of tectonic plates, or the proximity to volcanoes, because the power and scope of such energy is obvious when an earthquake or other crustal movement occurs. But a map of potential geothermal locations is not a static thing, since the Earth's crustal plates move, sometime with little notice (since we continue to research their exact locations and their levels of activity).

As well, geothermal energy can be associated with the presence of geysers, since these are spots in which a huge amount of energy, usually in the form of steam, is vented to the surface of the Earth. Famous geysers, like those in Yellowstone Park, in the United States, are preserved for their beauty and natural wonder. But lesser known geysers can be utilized for power generation. For instance, in northern California, the Mayacamas Mountains produce steam from roughly 350 wells, and have been harnessed in what is now the largest steam field reservoir in the world [10].

https://doi.org/10.1515/9783110662276-007

Table 7.1 shows a non-exhaustive list of geothermal power plants, including their locations.

Table 7.1: Geothermal energy locations [11].

Name	Location	Output (MW)	Comments
CalEnergy Generation's Salton Sea Geothermal Plant	California, USA	340	
Cerro Prieto Geothermal Power Station	Mexico	720	
Darajat Power Station	Indonesia	259	
The Geysers Geothermal Complex	California, USA	900	Currently, world's largest
Hellisheidi Geothermal Power Plant	Iceland	303	
Larderello Geothermal Complex	Italy	769	First plant of the complex built in 1913
Makban Geothermal Complex	Laguna and Batangas	458	
Malitbog Geothermal Power Station	Philippines	232.5	
Tiwi Geothermal Complex	Philippines	289	
Wayang Windu Geothermal Power Plant	West Java, Indonesia	227	

It is noteworthy that the locations in Table 7.1 correspond well to areas of the Earth that are prone to earthquakes, or that have active volcanoes. Several of them are located along the "ring of fire," more properly called the Circum-Pacific Belt.

Although Iceland does not dominate the list in Table 7.1, it has become something of a model country for the generation of power via geothermal energy. The reason is at least twofold. First, chemically, because the island nation is essentially the highest point of the Mid-Atlantic Ridge, it has an abundance of easy-to-reach, deep hot zones with molten or near-molten rock. This can be exploited in terms of using that heat to heat water, turn turbines, and ultimately to generate electricity. Second, politically, the nation is small and the government is not particularly complex; and thus, decisions about the use and exploitation of the nation's heat resources do not tend to have to go through numerous layers of government examination and approval before some decision is made. Approximately 25% of the nation's electricity comes from geothermal plants. Table 7.2 lists several of the major plants operating in Iceland.

Table 7.2: Geothermal power plants in Iceland.

Name	Output (MW)	Comments
Hellisheidi Power Station	303	Supplying power to Alcoa plant for aluminum refining
Karahnjukar Hydropower Station	690	Produces electricity for Alcoa operations
Krafla Power Station	60	
Nesjavellir Geothermal Power Station	120	
Peistarreykir Power Station	90	
Reykjanes Power Station	100	
Svartsengi Power Station	76.5	

Notes have been given in Table 7.2 of the use of geothermal power to produce elemental aluminum. This is because of all the metals, aluminum is one of the most energy intensive to extract from its major ore, bauxite. There have been both positive and negative implications to this use of Iceland's power, depending upon the source. It has been lauded as changing the world's economy through the inexpensive production of an industrially useful metal. It has also been attacked because of the environmental degradation and damage done to the local environments where the metal is refined.

7.3 Types of geothermal power plants

Geothermal power plants can be divided into three basic types:
1. Dry steam power plants
2. Flash steam power plants
3. Binary cycle power plants

Dry steam plants are the oldest design, while binary cycle plants are the newest. The three represent what can be called an evolution in plant design. The evolution has been aimed at producing power using progressively lower temperatures.

The concept of a geothermal power plant as an injection well or series of wells is fairly straightforward. A liquid, usually water, is injected into some aperture or fissure in the Earth, where it is exposed to magma or some other hot, underground zone, which can be solid rock. The resulting heated water is captured for use. This takes place by heating the water, routinely to steam, then using it to turn a turbine to generate electricity. It should be noted that some wells do not need to have water injected, because it exists in what is called the reservoir. Figure 7.1 shows a basic schematic of such a design.

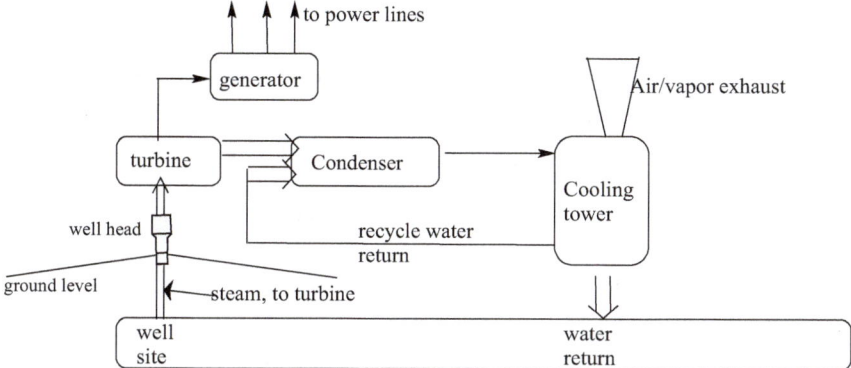

Figure 7.1: Basic geothermal power plant design.

7.3.1 Dry steam power plants

The design for a dry steam power plant is the simplest of the three, and as might be expected, this is the most mature design. The steam is either from a below-ground reservoir, or from water that is introduced, and generally results in steam, that is, at 150–180 °C. It is the force of the steam exiting the well to the surface, through what is generally called a well head that turns turbines.

Even in a dry steam plant, the steam is cooled and returned to the well, which is the purpose of the condenser. Returning the cooled water to the below-ground site makes the entire process a loop and recycles the water. At times, water must be added to such loops, since even closed loops have occasional escapes of steam.

7.3.2 Flash steam power plants

Flash steam power plants have become the predominant design for geothermal plants, and function by using water at 180 °C or higher, and flashing this high-temperature, high-pressure water to steam in what is called a separator before the fluid reaches the turbines.

The design schematic shown in Figure 7.1 would be almost the same for a flash steam plant as what is shown. A separator would be added between the well head and the turbine, and water would flow both from that separator to the return water line, as well as from a cooling tower/condenser to the return water line.

7.3.3 Binary cycle power plants

The youngest of the power plant designs is that of a binary cycle plant. This name indicates that two fluids are involved in getting to the point at which turbines are turned. Figure 7.2 shows the basic design of a binary cycle plant, and the steps for which are also listed here:
1. Water is pumped to the surface from the geothermal well site.
2. The water must be pumped through a heat exchanger.
3. The cooled water, after being passed through the heat exchanger, is returned to the geothermal site.
4. A low boiling fluid is gasified through the heat exchanger. Fluids that have been used successfully include C4 or C5 hydrocarbons.
5. The gasified low boiling point fluid then turns turbine blades.
6. The gasified fluid is then recondensed.
7. The liquid is then recycled to the heat exchange chamber.

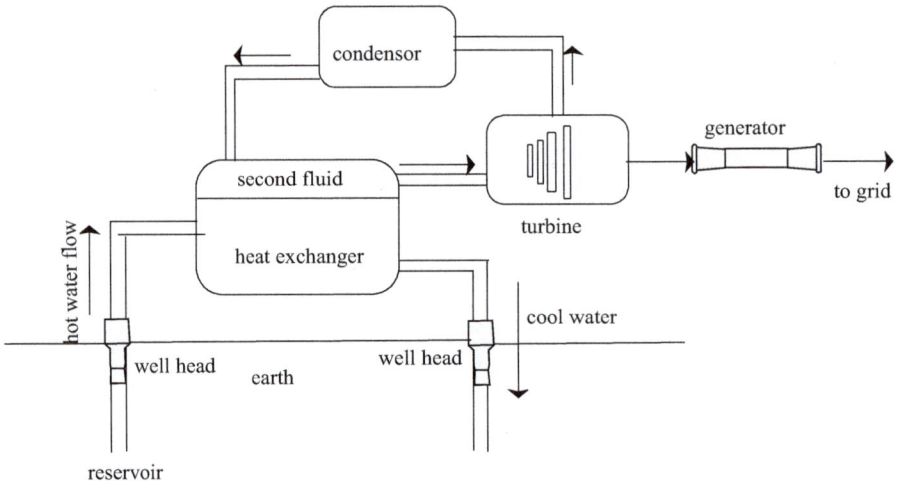

Figure 7.2: Schematic of binary cycle geothermal plant.

One aspect of the binary plant design that is not obvious in Figure 7.2 is that the second fluid is the one with a significantly lower boiling point than that of water. Because of this, binary plants can function with below groundwater that is of lower temperature than the dry steam or flash steam power plants. This becomes an important factor in determining the location of, and construction of, new plants.

A common factor in all the above designs is that water can be recycled after use at the turbine or heat exchanger, and then reintroduced to the earth and the strata from which it came.

7.4 Residential geothermal energy

There is another aspect to geothermal energy that is far removed from the big projects which generate electricity for large areas and large population concentrations. What might be considered "home geothermal" is a means whereby an individual house can be heated or cooled using water, in an economically more feasible method than traditional heating and air conditioning alone. A heat pump in the home helps it to function as an HVAC system. It should also be noted that this does not have to be limited to an individual residential home. It can be applied to commercial buildings and other structures.

7.4.1 Heat pumps

Heat pumps do not have to be large, industrial-scale plants for the production of electricity for some large area, utilizing deep shafts and extremes of temperature. They can be small units built in for use in a single home. Such units take advantage of the relatively stable temperature that exists in the earth several meters below the bed of many houses (or in a pond, more often seen near rural homes). The steps of such a system are as follows:
1. In a loop, cool water is pumped out of the house, to a lower level in the ground.
2. The water is then heated through contact of the piping system with the earth.
3. Then brought back to the house.
4. A conventional heat pump is installed.
5. The heat in the returned water is exchanged with air that is in contact with the heat pump, and warm air is pumped through the building's duct work.

Figure 7.3 shows the basic scheme and movement of water. Note that other fluids can be used in such a loop, but water is often preferred because it is inexpensive and poses no pollution problems should the system leak [12].

There is a growing movement to construct residences with what can be called this geothermal loop built into them, or rather, built partially under them. Using just one example of the benefits of such a feature to a house, the cost of heating water for use in showering or washing dishes can be dramatically lowered if the "cold" water entering is partially heated by having it flow through coils and pipes under the house, where the temperature of the ground is relatively constant, before it arrives at the tank of the hot water heater.

Once again, the chemistry in this environment remains a matter of heat exchange. In this case, water pumped to a house is warmed by being sent through the underground coils, before being stored in the traditional hot water tank.

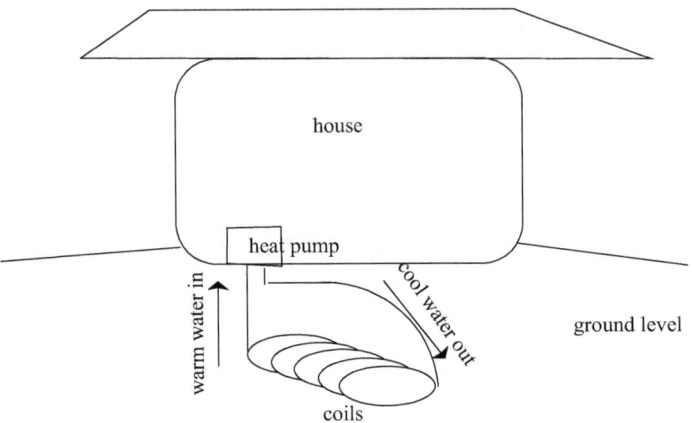

Figure 7.3: Water flow diagram of home heat pump.

References

[1]   Geothermal Energy Association. Website. (Accessed 20 May 2019, as: http://www.geo-energy.org/).

[2]   CanGEA, Canadian Geothermal Energy Association. Website. (Accessed 20 May 2019, as: https://www.cangea.ca/).

[3]   EGEC Geothermal. Website. (Accessed 20 May 2019, as: https://www.egec.org/).

[4]   International Geothermal Association. Website. (Accessed 20 May 2019, as: https://www.geothermal-energy.org/).

[5]   Geothermal Education Office. Website. (Accessed 25 July 2021, as: https://geothermaleducation.org).

[6]   Geothermal Heat Pump Consortium. Website. (Accessed 25 July 2021, as: https://geoexchange.org).

[7]   Geothermal Rising. (formerly: Geothermal Resources Council). Website. (Accessed 25 July 2021, as: https://geothermal.org).

[8]   The Geysers. Website. (Accessed 25 July 2021, as: https://geysers.com).

[9]   National Geothermal Collaborative. Website. (Accessed 25 July 2021, as: https://osti.gov/servlets/purl/883304).

[10] Website. (Accessed 26 July 2021, as: https://ww2.energy.ca.gov/title24/tour/geysers/index.html).

[11] Climate Care. Website. (Accessed 26 July 2021, as: https://climatecare.com).

[12] Office of Energy Efficiency & Renewable Energy: Building America Solution Center, Geothermal Heat Pumps. Website. (Accessed 26 July 2021, as: https://basc.pnnl.gov).

Chapter 8
Solar power

8.1 Solar power – introduction

The heat of the sun has been used by people for millennia for everything, from dry-
ing foods and leather goods to producing salt from brine ponds. People have recognized the sun's heat and power probably for as long as we have looked up to it, but
have for almost as long not been able to harness too much of that radiant power.
Indeed, when we compare the amount of energy provided by the sun to the Earth
each day, even today far more than 99% is not captured or used for power generation or in any other way at all.

Although the idea of using the sun's energy is ancient, the idea of solar voltaic
energy has a very short history, really starting only in 1946, with a patent by Russell
Ohl [1]. Likewise, harnessing what is now called concentrated solar power (CSP)
also has a relatively brief history, although an Italian patent for it was filed by Alessandro Battaglia in 1886. As well, one legendary use of it is rooted in ancient tales,
supposedly in an idea belonging to Archimedes.

Solar power falls into two broad categories: solar photovoltaics (PV), and CSP,
both of which we will discuss. The potential for solar to provide a very "green"
power footprint, meaning one that is environmentally sound and relatively low pol-
luting, is one aspect of it that is mentioned repeatedly by the many organizations
that exist to promote solar power [2–38]. Indeed, the opening message at the Solar
Energy Industries Association (SEIA) states:

> The Solar Energy Industries Association ® (SEIA) is leading the transformation to a clean en-
> ergy economy, creating the framework for solar to achieve 20% of U.S. electricity generation
> by 2030. [2]

Other organizations are similarly positive about the potential of solar energy, and
about its expansion in the near future.

8.2 Solar photovoltaics (PVs)

As the name implies, PV energy is that produced when the sun's light is absorbed
and transformed into electric energy. This is done with designed materials, all of
which often take advantage of silicon-based solids, some of which include small
amounts of an element like gallium, and some of which include small amounts of
elements like phosphorous or arsenic. These additions are called p-type doping and
n-type doping.

https://doi.org/10.1515/9783110662276-008

8.2.1 Photovoltaic function

Whatever the chemical makeup, PVs are a means of directly converting the sun's energy into electricity without having to turn a turbine, and thus power coupled to some type of electrical generator.

A stepwise description of what a solar cell is composed of and how it functions is as follows:

1. Solar cells are made of two semiconductor layers, generally silicon-based, called a p-type layer and an n-type layer.
2. The n-type layer is silicon doped with a small amount of an element such as phosphorus or arsenic which has an extra electron compared to silicon.
3. The p-type layer is silicon doped with a small amount of an element such as gallium, indium, or boron, which has one less electron than silicon. Such elements are said to create positively charged "holes."
4. The p and n semiconductor layers are placed next to each other, where electrons from the n-layer can move into the holes in the p-layer. Movement of electrons from the n-layer to the p-layer produce what is called a "depletion zone" with electrons filling holes.
5. When the depletion zone is filled, the n-layer will possess cations (positive charges), while the p-layer will possess anions (negative charges) – which makes an electric field.
6. Sunlight irradiating the PV material will eject electrons from the silicon, creating holes – which will then shift the electrons from the p-layer to the n-layer, re-creating holes in the p-type layer.
7. This release of electrons because of the sun's shining on the material is an example of the photoelectron effect.
8. Connecting the p-layer and n-layer with a conductor, such as a metal wire, electrons move from the n-layer to the p-layer and return to the n-layer via the wire. This is an electrical flow or current.

Figure 8.1 shows the basic layers of a solar cell, and how it connects to some end point that receives electric energy from them.

PV solar cells can be linked in large arrays or small. Thus, they can be used to produce large amounts of electricity for a single building, or to be part of the electric grid. They can also power a small object. This latter function is especially useful when some item requires power but is far from the existing electric power lines. Figure 8.2 shows one large array with acres of linked PV solar cells.

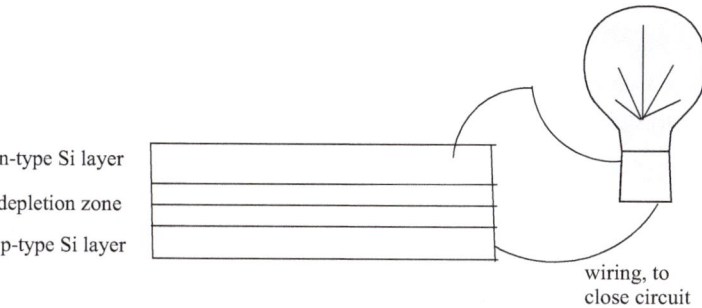

n-type Si layer

depletion zone

p-type Si layer

wiring, to
close circuit

Figure 8.1: Solar cell components.

Figure 8.2: Large-scale solar cell array.
(Photo, public domain, courtesy of National Renewable Energy Laboratory)

8.2.2 Photovoltaic material sources

While the "cleanness" of solar energy and PV cells is promoted heavily, both by the industries involved in their production as well as by other enthusiasts, a brief note should be made about the environmental cost of producing these materials [39].

Silicon

High-purity silicon of the level needed for PV production is a very small portion of the silicon used annually, the majority being used in metal alloys. But what is called a zone refining furnace is employed to refine silicon of the highest purity. This is an extremely energy-intensive process.

Gallium

All gallium is produced as a by-product of aluminum manufacturing or of zinc refining. The tailings from bauxite production, a strongly basic solution of sodium hydroxide, must have gallium extracted from it.

Arsenic

High-purity arsenic is a small use of all arsenic produced annually but is routinely made into gallium-arsenide for use in PVs. Since arsenic is found in nature in several different minerals, it must be extracted and purified from them.

Phosphorus

Phosphorus is always found in phosphate minerals, and is indeed tracked nationally in that form. Extraction of it from such minerals often involves purifying a compound such as calcium phosphate, then reducing the phosphorus with a reducing agent such as elemental carbon, thus coproducing carbon monoxide.

The economic and environmental costs of producing PVs does not make them less attractive as a source of energy, especially if they are designed to last for decades, which will mean pollution-free power for all that time. But the needs of such refining of elements should still be factored into any equation relating to PV energy production.

8.3 Concentrated solar power (CSP)

The most ancient use of solar power concentrated by some means is the stuff of legends. As mentioned above, supposedly, the ancient Greek genius Archimedes conceived of, and had built, several metal mirrors that could be adjusted, and that focused the sun's rays to a specific point. This focused solar power was then trained at enemy ships, and eventually set them on fire. While much has been written about this, often in very colorful language, virtually all scientists and engineers who have examined the phenomenon in modern times consider this a legend that has grown with the retelling, and not a device that was ever based in fact.

In a much more modern setting, the production of energy through the use of the sun to heat some material that then moves or in some way turns a turbine connected to a generator, and thus produces electrical energy, is the basis of what is called CSP. As one might imagine, working CSP plants tend to be located in desert or other sunny areas, where sunshine is common for most days of the year.

CSP is an evolving means by which electricity is generated, and therefore, there are several different types of designs. The four major ones are treated below. As well, what is called household solar is discussed because there is an enormous potential still to be realized from using some form of solar heating that is very delocalized, one house at a time.

Table 8.1 shows a non-exhaustive list of CSP plants throughout the world, generally listing the largest in any particularly country. Some countries, such as Spain, the United States, and Israel, have several such plants and are continuing to build more. The designs shown in the third column are all discussed in Sections 8.3.1 to 8.3.4.

Table 8.1: Examples of CSP plants worldwide.

Name	Location	Output (MW)	Design	Comments
Ashalim Power Station	Israel	120	Parabolic trough	
Dhursar Power Station	India	125	Fresnel lenses	
Ivanpah Solar Power Facility	California, USA	390	Solar power tower	Online since 2014
Kathu Solar Park	Northern Cape, South Africa	100	Parabolic trough	Kathusolarpark.co.za
Noor/Ouarzazate Solar Power Station	Morocco	510	Solar power tower and parabolic trough	World's largest
SEGS – Solar Energy Generating System	Mojave Desert, USA	310	Parabolic trough	
Solaben Solar Power Station	Spain	200	Parabolic trough	On-line since 2012

8.3.1 Fresnel reflectors

The idea of using numerous flat mirrors that are all in line to reflect sunlight to tubes filled with some fluid is the principle behind Fresnel reflectors. Much like the parabolic troughs to be discussed below, Fresnel reflectors are used to heat water or another liquid being passed through a tube. The heated water can then be depressurized, and the resulting gas – often steam – can be used to turn turbines connected to generators. If some other fluid is used, it is brought into contact with water at high temperature, and the water vaporized so that steam can turn turbines.

8.3.2 Parabolic troughs

The idea of using long tubes with fluid in them – often a molten salt – focusing mirrors around the tube as a central point, and heating fluid at that point is the same with both Fresnel reflectors and parabolic troughs. The difference is the shape and size of the mirrors. Figure 8.3 shows an array of parabolic trough collectors. Once again, the heated, pressurized fluid is put in contact with water to be vaporized to turn turbines which are connected to generators. The generators are in turn connected to an existing power grid. Figure 8.3 shows an example of parabolic troughs in a linear array.

In comparing parabolic troughs with Fresnel reflectors, it should be noted that the parabolic shape is more effective at concentrating sunlight to the central, fluid-filled tube. However, the cost of the Fresnel reflector design is the lesser of the two.

Figure 8.3: Parabolic troughs for CSP.
(Photo, public domain, courtesy of National Renewable Energy Laboratory)

8.3.3 Solar power tower

The word "tower" in this term is the central idea behind this form of CSP. Mirrors can be positioned in a circle around a central tower, and the sun's rays focused on some point of the tower. Often termed "heliostats," these mirrors can be turned via computer control to maximize the angle at which they catch sunlight. The focal point or receiver is then hot enough to heat a molten salt, which can then be brought into contact with water to create steam, which again turns turbines. Figure 8.4 shows one such arrangement. For a sense of size, note the trucks parked within the inner ring of heliostats.

Figure 8.4: Solar power tower CSP.
(Photo, public domain, courtesy of the National Renewable Energy Laboratory)

8.3.4 Stirling engine

The idea of a Stirling engine is not new, having been invented in the early 1800s, but its use as a means of transferring the power of expanding and contracting gas to a turbine is dating only to the 1970s. The idea is that a parabolic dish concentrates the sun's energy to a specific point, then the heat expands a gas in a cylinder, which can again contract. In this process, a wheel or turbine is turned. Figure 8.5 shows the basic schematic of a Stirling engine. Figure 8.6 shows an entire parabolic dish and Stirling engine.

In the Stirling engine, the gas-filled cylinder has an internal displacer that is slightly smaller than the diameter of the cylinder, but that pushes gas enough to push the piston, attached to the top of the cylinder. The pivoting arm attached to both the piston and the turbine provides the movement of the turbine, which is once again connected to a generator.

In Figure 8.6, note the relative size of the mirrors on the right with the Stirling engine and focal point on the upper left.

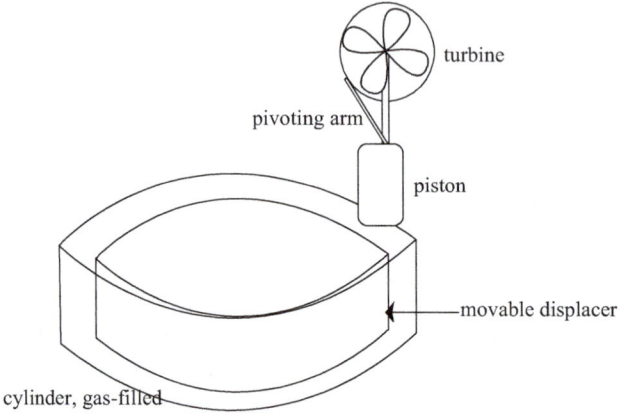

Figure 8.5: Stirling engine diagram.

Figure 8.6: Stirling engine CSP.
(Photo, public domain, courtesy of the National Renewable Energy Laboratory)

Although the field of CSP continues to expand, the chemistry of the materials which make the reflective surfaces tend to be mature ones. Metals can be used as reflective surfaces, as can silver-coated black-backed glasses.

8.3.5 Household CSP

One other form of CSP that is very often used by individual households is that of hoses and a flat surface oriented toward the sun, used to heat swimming pools or household water. Online sources exist, showing how to make such devices using parts that are as simple as plywood, garden hoses, and plastic or glass window coverings. By exposing water that is flowing through a hose which has been attached to a background that is painted black, and covering the hose and background with a plastic or glass covering, water piped from a house can easily be raised to 30–35 ° C from its starting temperature. Figure 8.7 shows a schematic of how a garden hose can be used to increase the temperature of water for a pool or household.

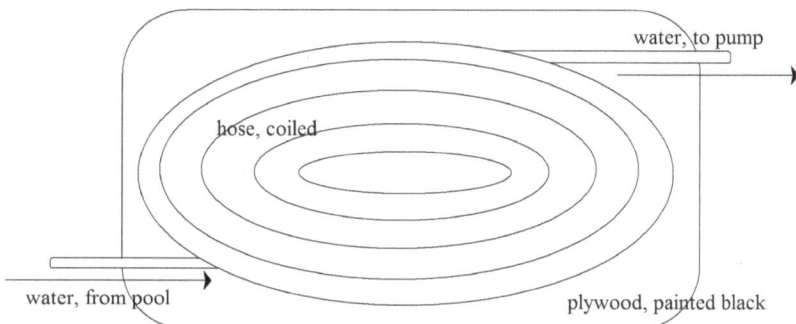

Figure 8.7: Home water-based CSP.

If this seems outlandish, it should be kept in mind that these small, household versions of a swimming pool heater or household water supply heater can be installed in many homes by DIY persons – do-it-yourself-ers. They do not cost a great deal and are easy to install even after a home has been built, meaning it is not something that must be built along with the structure.

8.4 Thermal gradient – wind/solar updraft towers

It is possible to derive electrical energy from the simple temperature gradient of air heated by the sun to different temperatures at different heights. Desert areas, while hot where the air meets the Earth, can be much cooler at higher elevations. This temperature difference has been the basis of several past energy projects, often called wind towers, solar towers, or solar chimneys. The general idea is shown in Figure 8.8.

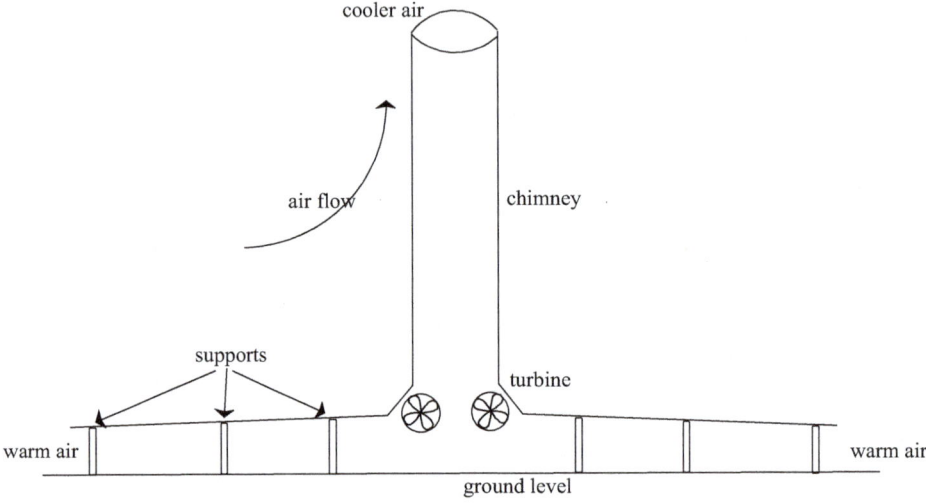

Figure 8.8: Solar updraft tower schematic.

It is noteworthy that any such towers that have been constructed must cover acres of land at the base. Also, the selection of materials in what can be called the cone over the base is important in that it helps heat the air in the immediate vicinity. The placement of turbines is such that they catch the updraft of the warm air through the tower; thus, they are generally located at the base. It is the turning of the turbines by moving air that is used to generate electrical power in such towers.

Solar towers have been considered as a possibility for power generation for decades but the idea has not caught on as readily as some other methods. One such plant in Manzanares, Spain, functioned for several years, but was eventually shut down. The chimney had a height of almost 200 m, which provided sufficient change in temperature from the base to the apex that turbines could be turned by the rising air [40].

References

[1]   Light Sensitive Device. Russell Ohl. U.S. Patent 2,402,662.
[2]   Solar Energy Industries Association. Website. (Accessed 26 July 2021, as: https://www.seia.org/).
[3]   Canadian Solar Industries Association. Website. (Accessed 26 July 2021, as).
[4]   Solar Energy UK. Website. (Accessed 26 July 2021, as: https://solarenergyuk.org).

[5]   Solar Power Europe – Leading the Energy Transition. Website. (Accessed 26 July 2021, as: https://www.solarpowereurope.org).

[6]   Associations > Solar. Website. (Accessed 26 July 2021, as: https://www.energy.eu).

[7]   Japan Photovoltaic Energy Association. Website. (Accessed 26 July 2021, as: http://www.jpea.gr.jp).

[8]   The Australian Solar Council. Website. (Accessed 26 July 2021, as: https://www.energymatters.com.au).

[9]   SEANZ: Sustainable Electricity Association of New Zealand. Website. (Accessed 26 July 2021, as: https://www.seanz.org.nz).

[10] NSEFI – National Solar Energy Federation of India. Website. (Accessed 26 July 2021, as: https://www.nsefi.in).

[11] SAPVIA – The South African Photovoltaic Industry Association. Website. (Accessed 26 July 2021, as: https://www.sapvia.co.za).

[12] Mexico National Solar Energy Association. Website. (Accessed 26 July 2021, as: https://openei.org).

[13] SunPower. Website. (Accessed 26 July 2021, as: https://us.sunpower.com/).

[14] First Solar. Website. (Accessed 26 July 2021, as: http://www.firstsolar.com/).

[15] SunEdison. Website. (Accessed 26 July 2021, as: http://www.sunedison.com/).

[16] Sullivan Solar Power. Website. (Accessed 26 July 2021, as: https://www.sullivansolarpower.com/).

[17] American Solar Energy Society. Website. (Accessed 26 July 2021, as: https://www.ases.org/).

[18] IREC and The Solar Foundation. Website. (Accessed 26 July 2021, as: https://www.thesolarfoundation.org/).

[19] Bright Green Energy Foundation. Website. (Accessed 26 July 2021, as: http://www.greenenergybd.com/site/index.php).

[20] Abellon Clean Energy. Website. (Accessed 26 July 2021, as: http://www.abelloncleanenergy.com/).

[21] Orb Energy. Website. (Accessed 26 July 2021, as: http://orbenergy.com/).

[22] M-KOPA Solar. Website. (Accessed 26 July 2021, as: http://www.m-kopa.com/).

[23] The Solar Foundation – Advancing Solar Energy Use – Nonprofit. Website. (Accessed 26 July 2021, as: https://www.thesolarfoundation.org/).

[24] Smart Electric Power Alliance | SEPA. Website. (Accessed 26 July 2021, as: https://sepapower.org/).

[25] International Solar Energy Society. Website. (Accessed 26 July 2021, as: https://www.ises.org/).

[26] Solar | Student Energy. Website. (Accessed 26 July 2021, as: https://www.studentenergy.org/topics/solar).

[27] Increasing Energy Efficiency: Residential Solar power. Website. (Accessed 26 July 2021, as: https://www.asla.org/residentialsolar.aspx).

[28] Solar Energy International. Website. (Accessed 26 July 2021, as: https://www.solarenergy.org/).

[29] American Solar Energy Society. Website. (Accessed 26 July 2021, as: https://www.ases.org/).

[30] Inside Climate News. Website. (Accessed 26 July 2021, as: https://insideclimatenews.org/).

[31] Engaging Students in Solar Energy Deployment on Campus. Website. (Accessed 26 July 2021, as: www.secondnature.org/2016/06/13/engaging-students-solar-energy-deployment-campus/).

[32] Solar Schools | Generation 180. Website. (Accessed 26 July 2021, as: https://www.generation180.org/solar-schools).

[33] Student Driven Solar Can Power up Michigan Schools | Ecology Center. Website. (Accessed 26 July 2021, as: https://www.ecocenter.org/student-driven-solar-can-power-michigan-schools).

[34] Solar Power Europe. Website. (Accessed 26 July 2021, as: http://www.solarpowereurope.org/home/).

[35] International Technology Roadmap for Photovoltaic. Website. (Accessed 26 July 2021, as: https://itrpv.org).

[36] Florida Solar Energy Center. Website. (Accessed 26 July 2021, as: https://energyresearch.ucf.edu).

[37] Arizona Solar Energy Industries Association (AriSEIA). Website. (Accessed 27 July 2021, as: ariseia.org).

[38] California Solar Energy Industries Association, CALSEIA. Website. (Accessed 27 July 2021, as: https://www.calseia.org).

[39] U.S. Department of the Interior, U.S.G.S. Mineral Commodity Summaries, 2021, downloadable at: https://pubs.er.usgs.gov>mcs2021.

[40] Schlaich Bergmermann, SBP. Website. (Accessed 27 July 2021, as: https://sbp.de/en/about-us).

Chapter 9
Wind power

9.1 Introduction

Mankind has harnessed the wind for thousands of years, routinely in the form of sailing ships and windmills (also referred to as gristmills), and to a lesser degree and more recently for moving balloons and dirigibles. The use of sails has always been to move ships, from the relatively small ones of ancient times, ultimately to those of very large size, the clipper ships that were the pride of their companies, and indeed of countries, in the eighteenth and nineteenth centuries. The sails of wind vanes on windmills have been used for slightly over a millennium both to raise and move water – if we think of the small, traditional windmills on farms – as well as to move stone wheels in grist mills that grind meal. The first wind-powered grist mills appear to have been used in southwest Asia. Ultimately, this second traditional use has been translated into the use of wind to produce electricity, with the first windmill generator dating back to the late 1880s, and the work of Scottish engineer James Blyth.

Wind power today produces approximately 2% of the world's energy [1–4] with differing amounts in different countries. For example, Denmark derives approximately 20% of its electrical power from wind, yet is a small nation [5, 6]. Both China and the United States have a much smaller percentage of their power derived from wind, but are much larger nations.

The potential of electrical energy from wind is so great that there are numerous regional, national, and international associations dedicated to its development [1–25].

9.2 Geography

The siting of windmills and wind farms is a concern to all companies that generate electricity via wind power. Studies are performed prior to construction, both for planned onshore and offshore wind farms, although some areas where steady or constant winds are well known do not need particularly detailed analyses [5–7]. The chief concern in such studies is how often the wind blows with sufficient strength to turn the blades of wind turbines. Detractors of wind power incessantly point out that the wind does *not* blow all the time in any one place – although a legend or joke among people who live on the Orkney Islands north of Britain says that one day the wind stopped blowing, and everyone fell over at once!

Other factors in siting wind farms are not necessarily concerned with the science or engineering of a particular project but involve political and social considerations. For example, does the local and regional government want to project the

https://doi.org/10.1515/9783110662276-009

image of the wide use of renewable sources of energy? If so, siting a wind farm in their region becomes important. As well, do the residents of the local community consider the presence of these structures an advantage or a disadvantage for their community? This can also influence whether or not the investment will be made to build and maintain a wind farm.

Wind farms can be sited on land or on the open ocean. Table 9.1 is a nonexhaustive list of the world's largest offshore wind farms. Even offshore wind farms have factors that need to be examined, besides how much the wind blows in a particular area of the ocean. Use of a specific area of the ocean, either for marine traffic or for fishing, as well as national ownership of that part of the ocean, also needs to be considered when such a farm is to be built.

Table 9.1: Large offshore wind farms.

Name	Location	Capacity	Comments
Hornsea One	Irish Sea	1.2 GW	Almost complete, 2019
London Array	Thames, England	630 MW	Developed by Masdar, E.ON, and DONG Energy
Greater Gabbard	Suffolk, North Sea	500 MW	140 turbines, 3.6 MW each
Bard Offshore 1	North Sea	400 MW	80 turbines
Anholt	Denmark	400 MW	Majority Owned by DONG Energy
Walney	Irish Sea	367 MW	DONG Energy operator
Thorntonbank	North Sea, Belgium	325 MW	54 turbines
Sheringham Shoal	Norfolk, UK	316 MW	88 turbines
Thanet	Kent, UK	300 MW	
Centrica Links	Lincolnshire, UK	270 MW	75 turbines
Horns Rev 2	Denmark	209 MW	DONG Energy-owned

9.2.1 Wind farms

Since wind does not blow steadily throughout the planet, the siting for windmill generator placement – where wind farms will be built – becomes an important factor in the ultimate economic success of such ventures, as mentioned. Some areas have relatively consistent winds, areas such as Minnesota and northern Scotland, where wind power can be economically very profitable. Other areas must be studied to determine if there are enough windy days that construction of a wind farm in the area will be profitable, and over what span of time. One such area that has proven profitable in the past decade is southern Ontario, Canada, north of Lake Erie. Figure 9.1 shows some of the windmills typical of the area.

Another factor that must be considered in siting a wind farm is how far the farm will be from an existing power grid. Any electricity generated at the farm needs to be connected directly to the grid, or in some way stored for future use. Remote areas may meet all the requirements for a profitable wind farm but still be far enough away from the existing power grid that this one factor becomes a problem.

Figure 9.1: Examples of windmills on a modern wind farm.

Wind power in the United States has spread to 41 of the states and continues to grow. A list of some of the largest wind farms within the United States includes those listed in Table 9.2. A full list of all the operating wind farms in the United States would now be pages long.

Notice that even though wind farms exist in several states, there is an abundance of them in the Great Plains. This is noteworthy because the traditional view of a state like Texas is that it is "oil country." But companies have also embraced the potential of wind as an economically feasible way by which electric power can be generated.

Table 9.2: Wind farms in the United States.

Name	Location	Output (MW)	Comments
Alta Wind Energy Center, aka. Mojave Wind Farm	Kern County, California	1,550	Turbines supplied by Danish company, Vestas
Horse Hollow Wind Energy Center	Taylor and Nolan County, Texas	735	
Javelina Wind Energy Center	Texas	749	
Los Vientos Wind Farm	South-central Texas	910	Continues to expand
Meadow Lake Wind Farm	Indiana	801	
Peñascal Wind Farm	Texas	605	
Roscoe Wind Farm	Abilene, Texas	781	
Rush Creek Wind Project	Eastern Colorado	600	Operated by Xcel Energy, also deals in solar power
San Gorgonio Pass Wind Farm	California	619	
Shepherds Flat Wind Farm	Arlington, Oregon	845	Operating since 2012
Tehachapi Pass Wind Farm	Kern County, California	710	~3,400 turbines, began in the 1980s

9.2.2 Offshore wind power

The idea of siting wind farms offshore has several factors in common with those on land, and some that are unique to a marine environment. For example, as mentioned, the amount of other activity in the area must be considered, whether it is marine-based traffic or land-based traffic at a shore that interacts with marine-based traffic and activities. In the case of offshore farms, though, there is usually no concern for the visual impact of the operation – a consistent concern when the wind farm, and the shadow of the towers and the moving blades, falls on people's homes.

Table 9.1 shows a non-exhaustive listing of some of the world's largest offshore wind farms. It can be seen that one company owns or has interests in a significant share of the wind farms listed in Table 9.1. DONG Energy, renamed Ørsted A/S in 2017, was originally founded in 2006. This makes the company a very mature one in terms of working in wind-generated energy. But its history goes back several more decades, as a natural gas company. The name DONG is an acronym of "Danske Olie og Naturgas A/S." Since the nation of Denmark is essentially the Jutland peninsula and numerous islands, and the peninsula is seldom wider than 150 km, wind from the North Sea and Baltic Sea is fairly steady, making the nation ideally suited for the development of wind power.

In the United States, the only currently operational offshore wind farm is the Block Island Wind Farm located off the coast of Rhode Island. It can produce 30 MW of electricity from its five turbines, and began operations in late 2016. Estimates from the United States Department of Energy however indicate that the potential for offshore wind energy production to be orders of magnitude greater than this, simply because the nation has such long coastlines. As well, in 2010 the U.S. Department of the Interior established the Bureau of Ocean Energy Management within it, in part to manage development of renewable energy sources, including wind power. The website states, in part:

> The President announced on April 22, 2009, that the Interior Department completed the Final Renewable Energy Framework or rule making process to govern management of the Renewable Energy Program. The final rule establishes a program to grant leases, easements, and rights-of-way for orderly, safe and environmentally responsible renewable energy development activities, such as the siting and construction of offshore wind farms on the OCS as well as other forms of renewable energy such as wave, current and solar. [26]

The OCS mentioned above is the Outer Continental Shelf. Thus, this agency has a wide mandate to oversee development of wind energy off the shores of the United States.

9.3 Turbine designs

There are several different wind turbine designs which harness wind to generate electricity. Perhaps, the most basic and commonly adopted is that shown in Figure 9.2. This is the basic scheme for the type of wind turbine shown in Figure 9.1. Not shown in Figure 9.2 is a maintenance ladder, which can be on the outside or inside of the tower, and an anemometer, which measures wind speed and is located on or near the nacelle, and thus close to where the rotors are.

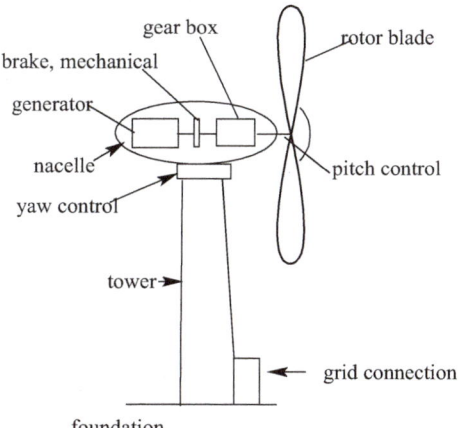

Figure 9.2: Wind turbine design.

Because the blades of a wind turbine can move in three dimensions, as can the entire nacelle, three terms "pitch, roll, and yaw" should be defined.

Pitch: rotation of the rotor up or down.
Roll: rotation about an axis from the rotor to the back of the nacelle.
Yaw: movement about the vertical axis of the tower.

Pitch, roll, and yaw control are all necessary because wind direction is not steady, and the assembly must be adjusted at times to best capture the wind. The size of rotor blades can vary greatly, with some of them being tens of meters long.

The materials used to make the gear box, brake, and generator – routinely various metals – are a constant concern for companies that manufacture these components. Some are lanthanides, and thus are scarce as the world markets for them change over time [27, 28]. For example, the gear box in such motors routinely uses neodymium–iron–boron magnets, which are some of the strongest known. The concern in the United States over lanthanide elements (also still called rare earth elements) such as neodymium is that the United States imports them, and thus there are governmental concerns over their continued availability. The USGS Mineral Commodity Summaries 2021 states:

> The U.S. Department of Defense announced contracts and agreements with rare-earth-element producers under the authorities of title III of the Defense Production Act. These agreements were put in place to support and strengthen the domestic rare earth supply chain in response to the Presidential Determinations. [28]

Lanthanides are not used solely in the motors of wind turbines, and thus must compete with other industries and user end items when it comes to their availability.

The materials used to make wind turbine blades – often fiberglass, carbon fibers, and balsa wood – are generally not considered harmful. Still, efforts are being made to use them or parts made from them in some further application when a turbine is decommissioned.

9.3.1 Vertical design

What can be called a traditional windmill, at least in the mind of the US population, is that of the windmill pump seen on farms throughout the American heartland. This now dated fixture of farming on the prairie has been displaced by the much larger wind turbines which are situated in close proximity, in wind farms, dedicated solely to the production of electricity. The traditional farm windmill was not used to generate electricity but rather to pump water.

Nevertheless, what most people consider and recognize as a modern windmill are those which have a vertical design, as shown in Figure 9.1: usually three large blades mounted on a single column. There are other vertical designs as well, however.

9.3.2 Darrieus turbines

Wind turbines that are placed in areas where space may be at a premium sometimes use what is called the Darrieus turbine design. These do not have individual blades, but rather a twisting design in which the surface that catches the wind surrounds the central shaft that arises from the ground. Considered a vertical axis wind turbine, the design is such that the wings or aerofoil blades can begin to rotate faster than the wind. The basic design is shown in Figure 9.3.

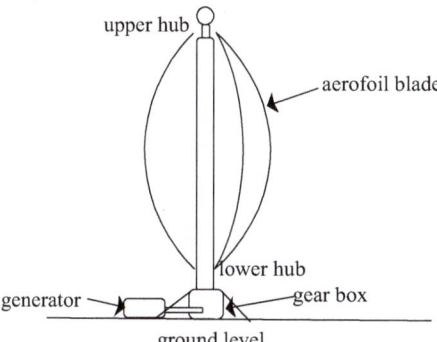

Figure 9.3: Darrieus wind turbine design.

9.3.3 Airborne designs

The idea of an airborne wind turbine is one in which the turbine has no hard element connecting it to the ground, like a tower. Such a design takes advantage of the fact that winds blow much more consistently at higher elevations, and thus can produce energy for greater periods of time during any day. Trials of such wind turbines have been undertaken in Alaska, in areas where the wind blows fairly constantly.

The design is much like a hot air balloon surrounding what looks like a fan. The firm Altaeros Energies has already tried what they term as "BAT" (Buoyant Airborne Turbine) at roughly 1,000 ft above the ground [29]. The outer "balloon" is helium filled, the fan is the center of the device, and guide wires tether the device to the ground as well as provide a platform for wiring that delivers the electricity. The trials were located in areas of Alaska where such devices would be able to provide electricity to remote towns and villages.

9.3.4 Others

It might be fair to say that there are as many designs for turbines and windmills as there are ways to catch the wind. The larger, more established ones are those mentioned above, but others exist, all of which generally fall into two broad categories:

vertical axis wind turbines or horizontal axis wind turbines. As mentioned, there have even been wind turbines made that exist as a form of high-altitude balloon or kite, one that stays airborne full time, while it generates electricity from catching the wind high above the Earth's surface.

9.4 Convective towers

Convective towers represent another way by which air flow can be harnessed. On a large scale, these can be used to generate electricity, while on a small scale – the size of an individual house – they are more a passive means of heating or cooling the building.

The idea of a convective tower is not complicated. It is a tower that funnels air at a high level down through a central column or "chimney" which has turbines mounted at the base of it, or that allows movement of air in an up-to-down direction. This is very much like a solar chimney, which was discussed in some detail in Chapter 8.

9.4.1 Downdraft tower

The idea of a downdraft tower is relatively simple, as it is essentially a tall, open cylinder of approximately 200 m high, with warmer air at the top, into which water is sprayed as a mist. The water cools the air, making it somewhat more dense, and as the air flows downward it turns turbines at the base, which are connected to generators [30–32]. Figure 9.4 shows the simplified schematic.

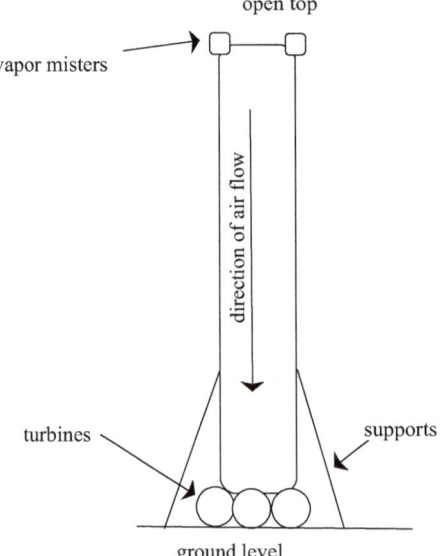

Figure 9.4: Downdraft tower design.

Recently, one company has designed and built wooden downdraft towers, in an effort to replace any steel that would be used in such devices, and in the process produces a tower that is itself more environmentally friendly than any previous design. An individual tower may produce only 1.5 MW of output, but can be made in large numbers, as with wind turbines or any other such devices.

9.4.2 Household solar chimney

The just-mentioned means of using wind power are all large-scale projects, yet there has been a significant amount of research and effort put into what might be called a household solar chimney [33, 34]. This design feature of buildings is not yet utilized universally and can have variations, all of which can still be called household solar chimneys. The idea appears poised to grow in the coming years. The design elements include the following:

1. Most basic is to paint the chimney black – thus moving air through the building. Energy saving is derived from heating or cooling that is not required by traditional HVAC. Other materials can be used besides paints [35].
2. An inlet into an at-ground or below-ground channel or tube must be present.
3. Warm air is pulled down the chimney through a chiller.
4. Warm air in the home is cycled out through vents in the ceiling and roof.
5. This becomes a form of passive solar heating.

More broadly, this is a form of passive temperature adjustment. Designs for this can differ from building to building.

9.5 Materials requirements

In all of these designs for vertical or horizontal wind turbines, as well as for cooling towers or chimneys, it can be seen that the cost and materials chemistry involved in harnessing wind for energy production is in the production of the wind turbine or other device, and the infrastructure involved in wiring the power sources into existing power grids.

As a recent example, the US President Joe Biden asked in a speech to the United States Congress in 2021 why the blades for wind turbines could not be made in the United States. While he spoke from the point of view of a politician, the question is based on the geographic locations of starting materials, specifically where the glass fiber or carbon fiber materials used in such blades is located. Similarly, the lanthanides previously mentioned, which are used in the motors of wind turbines, are currently sourced from only a few areas in the world, most notably China. Lanthanides have been produced in the western United States in the past, but their extraction is presently not economically feasible.

References

[1]   The American Clean Power Association. Website. (Accessed 27 July 2021, as: https://www. awea.org/).

[2]   The European Wind Energy Association. Website. (Accessed 27 July 2021, as: http://www. ewea.org/).

[3]   World Wind Energy Association. Website. (Accessed 27 July 2021, as: http://www.wwindea. org/).

[4]   Global Wind Energy Council. Website. (Accessed 27 July 2021, as: http://gwec.net/).

[5]   Danish Wind Industry Association. Website. (Accessed 27 July 2021, as: http://www.wind power.org/en).

[6]   Danish Wind Export Association. Website. (Accessed 27 July 2021, as: https://www.dwea.dk/).

[7]   New Zealand's Wind Farms. Website. (Accessed 27 July 2021, as: http://www.windenergy.org.nz/).

[8]   Canadian Wind Energy Association. Website. (Accessed 27 July 2021, as: https://canwea.ca/).

[9]   Polish Wind Energy Association. Website. (Accessed 27 July 2021, as: http://psew.pl/en/).

[10] Australian Wind Energy Association. Website. (Accessed 27 July 2021, as: https://tethys.pnnl. gov/institution/australian-wind-energy-association).

[11] Australia – Asia Wind Energy Association 2021. Website. (Accessed 27 July 2021, as: https:// www.asiawind.org).

[12] German Wind Energy Association. Website. (Accessed 27 July 2021, as: https://www.wind-energie.de/en).

[13] Japan Wind Power Association. Website. (Accessed 27 July 2021, as: http://jwpa.jp).

[14] Renewable UK. Website. (Accessed 27 July 2021, as: https://www.renewableuk.com).

[15] South African Wind Energy Association. Website. (Accessed 27 July 2021, as: https://sawea. org.za).

[16] Mexico Wind Power 2020 / Global Wind Energy Council. Website. (Accessed 27 July 2021, as: https://gwec.net>mexico-windpower-2020).

[17] Wind / REAP – Renewable Energy Alaska Project. Website. (Accessed 27 July 2021, as: https://alaskarenewableenergy.org).

[18] Montana Renewable Energy Association. Website. (Accessed 27 July 2021, as: https://monta narenewables.org).

[19] WINDExchange. Website. (Accessed 27 July 2021, as: https://windexchange.energy.gov/ states/nd).

[20] Great Plains Energy Corridor. Website. (Accessed 27 July 2021, as: https://www.energynd. com/resources/wind/).

[21] Women of Renewable Industries and Sustainable Energy. Website. (Accessed 27 July 2021, as: http://wrisenergy.org/).

[22] Blattner Energy. Website. (Accessed 27 July 2021, as: http://blattnerenergy.com/).

[23] Novatus Energy – Onward Energy. Website. (Accessed 27 July 2021, as: http://www.novatuse nergy.com/).

[24] Illinois Renewable Energy Association. Website. (Accessed 27 July 2021, as: https://www.illi noisrenew.org/).

[25] Advanced Power Alliance. Website. (Accessed 27 July 2021, as: http://windcoalition.org/).

[26] U.S. Department of the Interior, Bureau of Ocean Energy Management. Website. (Accessed 27 July 2021, as: boem.gov).

[27] Wind Power Monthly. Website. (Accessed 28 July 2021, as: http://windpowermonthly.com/article/1519221/rethinking-use-rare-earth-elements).

[28] USGS Mineral Commodity Summaries. 2021. Downloadable as: http://pubs.er.usgs.gov/publication/mcs2021.

[29] Altaeros. Website. (Accessed 28 July 2021, as: http://altaeros.com/products).

[30] New Atlas. Website. (Accessed 27 July 2021, as: http://newatlas.com/solar-wind-energy-downdraft-tower/32607/).

[31] New Atlas. Website. (Accessed 28 July 2021, as: http://newatlas.com/timbertower-wooden-wind-turbine/25007).

[32] N. Atlas. Website. (Accessed 28 July 2021, as: newatlas.com/energy/Sweden-first-wooden-wind-turbine-tower).

[33] A. Jafari, A.H. Poshtiri. Passive solar cooling of single-storey buildings by an adsorption chiller system combined with a solar chimney. Journal of Cleaner Production. 2017, 141, 662–682.

[34] X. Sun, Y. Sun, Z. Zhou, M.A. Alam, P. Bermel. Radiative sky cooling: Fundamental physics, materials, structures, and applications. Nanophotonics. 2017, Accessed as https://doi.org/10.1515/naoph-2017-0020.

[35] A. Sivanathan, Q. Dou, Y. Wang, Y. Li, J. Corker, Y. Zhou, M. Fan. Phase change materials for building construction: An overview of nano/micro-encapsulation. Nanotechnology Reviews. 2020, Accessed as https://doi.org/10.1515/ntrev-2020-0067.

Chapter 10
Energy storage

10.1 Energy storage – introduction

People wanted to store energy in some form for generations, almost since the moment we could generate it. The first practical use of what are now some means to store electricity goes back two centuries, to the mid-1700s. The devices that stored some amount of electrical charge are called Leyden jars, named after the city in which Pieter van Musschenbroek worked on such instruments in 1745. This type of jar was the first method or apparatus by which electrical charge could be stored for any length of time. Connecting several of them supposedly reminded some of the earliest people who experimented with them to make the comparison to batteries of cannon in a military formation. Supposedly it was Benjamin Franklin who named these "batteries."

Batteries are essentially a type of galvanic cell. This means it is a cell or couple in which a redox reaction occurs spontaneously. As we are familiar with, a redox reaction is any reaction in which one chemical species loses an electron or electrons (is oxidized) while another gains those electrons (is reduced).

Today, a wide variety of batteries – sometimes still called dry cell batteries – are produced by several companies and used in an enormous variety of consumer goods and products. Figure 10.1 shows a very small selection of the array of standard batteries which can be purchased in many grocery and household appliance stores.

10.2 Battery types

There are numerous different chemical couples that have been used to produce useful batteries, including those we now might think of as standard, as well as more exotic ones. Broadly, any battery that can be used only once, during its discharge cycle, is referred to as a *primary cell*. The discharge cycle may be years long, but if there is no possibility of recharging the battery, it is a primary cell. Any battery that can be recharged is referred to as a *secondary cell*. Batteries that are used to power small electrical devices come in a variety of shapes and power outputs, but have been standardized in the past decades. Table 10.1 shows a non-exhaustive list of them.

While the shapes of batteries vary widely, a few basic parts are always present. Figure 10.2 shows the basic makeup of a generic, cylindrical, dry-cell battery.

Another type of battery that has become standardized over the years, one that is not a small, cylindrical design, is the lead-acid battery used in most automobiles.

https://doi.org/10.1515/9783110662276-010

Figure 10.1: Examples of common battery types.

Table 10.1: Common types of batteries.

Type	Voltage (V)	Chemical couple, example	Common uses, example
AAA	1.5	Nickel–metal hydride, or NiCad	Small household electronics
AA	1.5	Nickel–metal hydride, or NiCad	Small household electronics
A	1.5	Nickel–metal hydride	Older laptops
C	1.5	Nickel–metal hydride, NiCad, or carbon–zinc	Flashlights, small electronics
D	1.5	Nickel–metal hydride, NiCad, or carbon–zinc	Flashlights, older radios
E, aka 9-volt	9	Nickel–metal hydride, NiCad, or carbon–zinc, lithium	Smoke alarms, radios
Lantern	6	Carbon–zinc	Lanterns
"Button"	Varies	Silver oxide, lithium	Cameras, wireless devices

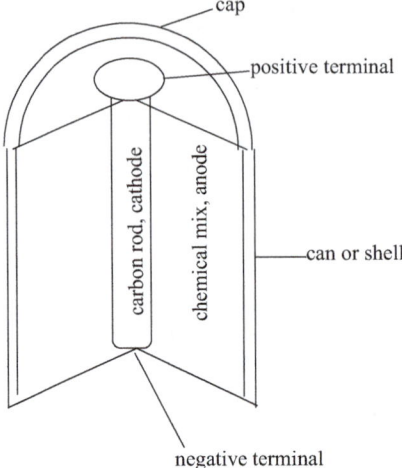

Figure 10.2: Basic components of a battery.

Figure 10.3 shows an example of one in an automobile. They are normally not used to power the vehicle, but to start them. This type of battery has become so ubiquitously used that there are now professional associations devoted to their use and recycling [1–31]. Older automobiles used a lead-acid battery to which distilled water was occasionally added. In the past 30 years, however, virtually all automobile batteries are self-contained, and do not require the user of the vehicle to add water to it.

Figure 10.3: Lead-acid automobile battery.

Notice that to the upper left of Figure 10.3 is a regulator (the red plastic cap has been opened to show the components underneath). This directs power to different parts of the automobile, as modern automobiles now often have computerized dashboard screens at the front, as well as DVD players or other screens mounted farther back.

Figure 10.4 gives some example of the variety of lead-acid batteries that are manufactured for different cars today. Roadside service providers such as the American Automobile Association keep multiple batteries in their service trucks to speed up roadside assistance.

Figure 10.4: Examples of different types of lead-acid batteries.

What all batteries have in common is some chemical couple – often two metals or metal salts – in which one chemical species is oxidized and another is reduced. In any primary cell, this reaction occurs irreversibly. In any secondary cell, the reaction can be reversed, returning the battery to its original state.

10.2.1 Lead-acid batteries

Many commercially available batteries function in a basic environment, but one old, firmly established battery functions in an acidic one, the lead-acid storage battery. Figure 10.3 shows a photo of one of the many that are currently manufactured for use in automobiles. Figure 10.5 shows the simplified reaction chemistry for it, first as a whole cell, and as half-cells.

$PbO_{2(s)} + Pb_{(s)} + 2H_2SO_{4(aq)} \rightarrow 2PbSO_{4(s)} + 2H_2O_{(l)}$

At discharge, the reducing half-cell (the cathode) is:
$PbO_{2(s)} + 3H^+_{(aq)} + HSO_4^-{}_{(aq)} + 2e^- \rightarrow PbSO_{4(s)} + 2H_2O_{(l)}$

...and the oxidizing half-cell (the anode) is:
$Pb_{(s)} + HSO_4^-{}_{(aq)} \rightarrow PbSO_{4(s)} + 2e^-$

Figure 10.5: Lead-acid battery.

These batteries now have more than a century of history to them. Few except historians remember that Ford Motor Company once marketed an electric car, one that ran completely off of lead-acid batteries. In truth, before the year 1913, Ford sold more battery-powered automobiles than it did gasoline-powered automobiles. Upon knowing this, some advocates of electric autos want to know why such technology cannot simply be reproduced. The answer generally has to do with what is expected of an automobile today in terms of speed and acceleration, and what was expected in 1913. In short, the consumer wants more now, wants exactly what can be provided by an internal combustion engine.

In that past time, not only was quick acceleration not as important as now, but individuals who had the means to afford one automobile usually had the money to afford two. Thus, an electric car was used in cities and towns where trips were short; and a gasoline-powered car was used for long trips. The only type of battery used in these electric cars was lead-acid batteries, simply because that was the state of the art at the time. But also at that time, cars did not accelerate from 0 to 60 mph (100 kph) in a few seconds, as they are expected today.

To achieve the voltage needed in an automobile, cells are connected in any lead-acid battery. The standard voltage for a single lead-acid battery is 2 V; therefore, six cells are linked in series to create 12 V that is common in a car battery.

10.2.2 Zinc-manganese dioxide cell

What has historically been called the Leclanché cell provides 1.5 V and has been used as a source of single-use power, since it is a primary cell. Figure 10.6 shows the chemistry for its discharge, both as a full cell and as half cells.

$Zn_{(s)} + 2MnO_{2(s)} \rightarrow ZnO_{(s)} + Mn_2O_{3(s)}$
Half cells:
$Zn_{(s)} + 2OH^-{}_{(aq)} \rightarrow ZnO_{(s)} + H_2O + 2e-$ anode
$2MnO_{2(s)} + H_2O + 2e- \rightarrow Mn_2O_{3(s)} + 2OH^-{}_{(aq)}$ cathode

Figure 10.6: Zinc-manganese cell.

In somewhat more detail, the anode of the cell is zinc, and the cathode is carbon rod (C_6), with a $C_6/MnO_2/NH_4Cl$ paste.

10.2.3 NiCad batteries

As given in Table 10.1, this battery couple sees extensive use in smaller electronics. This common couple produces a potential of about 1.4 V and is a useful secondary cell. Extensive research has been done on how many times such batteries can be discharged, then recharged – meaning how many cycles they can undergo and still be useful. Figure 10.7 shows the cell and half-cell reactions.

$Cd_{(s)} + NiO_{2(s)} + 2H_2O \rightarrow Ni(OH)_{2(s)} + Cd(OH)_{2(s)}$

As half cells:
$Cd_{(s)} + 2OH^-_{(aq)} \rightarrow Cd(OH)_{2(s)} + 2e-$ anode
$NiO_{2(s)} + 2H_2O_{(l)} + 2e- \rightarrow Ni(OH)_{2(s)} + 2OH^-_{(aq)}$ cathode   **Figure 10.7:** NiCad cell.

This battery couple remains widely used because it has relatively high energy density, meaning energy per unit volume.

10.2.4 Nickel–metal hydride batteries

As given in Table 10.1, nickel–metal hydride batteries are widely used in applications where a voltage must be supplied to some small piece of equipment. The couple gives 1.35 V and is capable of storing up to 50% more energy by volume than the NiCad battery. It too is a secondary cell. Figure 10.8 shows its reaction chemistry.

$MH_{(s)} + NiO(OH)_{(s)} \rightarrow Ni(OH)_{2(s)} + M_{(s)}$

As half cells:
$MH_{(s)} + OH-_{(aq)} \rightarrow M_{(s)} + H_2O + e-$ anode
$NiO(OH)_{(s)} + H_2O + e- \rightarrow Ni(OH)_{2(s)} + OH^-_{(aq)}$ cathode   **Figure 10.8:** Nickel–metal hydride cell.

The reason the name "nickel–metal hydride" is not more specific is that more than one alloy can be used as a couple with nickel. $LaNI_5$ and Mg_2Ni are both examples that have seen use.

10.2.5 Silver oxide batteries

Another battery couple with a long history to it is what is called the silver oxide battery, even though the two metals in it are silver and zinc. The couple has a relatively high voltage, 1.86 V, and has a shelf life that is years long, especially when stored in a cool but not frozen environment. Figure 10.9 shows the reaction chemistry for the couple.

$$Zn_{(s)} + Ag_2O_{(s)} \rightarrow ZnO_{(s)} + 2Ag_{(s)}$$
As half cells:
$$Ag_2O_{(s)} \rightarrow Ag_{(s)} \qquad \text{cathode}$$
$$Zn_{(s)} \rightarrow ZnO_{(s)} \qquad \text{anode}$$

Figure 10.9: Silver oxide battery.

An electrolyte of a KOH or NaOH solution is also required for this cell. Also, zinc can activate the electrolyte under certain conditions, causing H_2 production and an explosion. A very small amount of mercury suppresses this.

We will discuss lithium batteries, below, but it should be noted here that the silver oxide battery has approximately 40% more run time than Li^+ battery. It has lower energy density, however.

For the most part, silver oxide batteries are used in relatively small applications, and are often called button batteries, as mentioned in Table 10.1. But the most exotic application of specially made silver oxide batteries probably remains the lunar rover that astronauts used on the Moon in the 1970s. The specialized battery pack made for the rover was theoretically capable of traveling 51 miles (~90 km) on the Moon's surface. However, the rover stayed far closer to the lunar module. This was not because the silver oxide batteries were unreliable or might discharge more rapidly on the Moon's surface than they would have on the Earth. Rather, in case any component of the rover failed, the astronauts were instructed to stay close enough to the launch module that their backpacks and suits could carry them back to it at a bounce.

10.2.6 Mercury batteries

The use of elemental mercury and of mercury compounds has been phased out in many countries because of the high toxicity of mercury salts. However, before such materials were discontinued, what were called mercury batteries were used as a type of button battery, as mentioned in Table 10.1, and proved to have extremely long working lives – approximately 10 years, with a 1.35 V cell potential. The reaction chemistry for such batteries is shown in Figure 10.10.

In such batteries, often the anode was separated from the cathode simply with a layer of paper or other porous material soaked with the NaOH or KOH electrolyte. The mixed oxide shown as the cathode as well as the nonstoichiometric

$HgO_{(s)} + Zn_{(s)} \rightarrow Hg_{(l)} + ZnO_{(s)}$

As half cells:
$HgO(MnO_2)_{(s)} \rightarrow Hg_{(l)}$ w/ C_6 mixed in   cathode
$Zn_{(s)} \rightarrow ZnO_{(s)}$ anode   **Figure 10.10:** Mercury battery.

amount of graphite is so that mercury never accumulates into a large droplet. Additionally, a small amount of extra mercuric oxide is put into the cell to prevent evolution of hydrogen gas at the end of the cell's working life.

10.2.7 Lithium batteries

A great deal of effort has been put into research involved in producing long-lasting, quick-recharging batteries that utilize lithium. This is simply because lithium is the least dense metal on the periodic table, and therefore should result in lighter batteries for the same energy output. This becomes extremely important in the current and future production of electric automobiles and light trucks. Figure 10.11 shows the reaction chemistry for a lithium metal couple, while Figure 10.12 shows the reaction chemistry for what has become a common lithium ion battery.

$Li_{(s)} + MnO_{2(s)} \rightarrow MnO_2(Li^+)_{(s)}$

As half cells:
$MnO_{2(s)} + Li^+ + e- \rightarrow MnO_2(Li^+)$ cathode
$Li_{(s)} \rightarrow Li^+ + e-$ anode   **Figure 10.11:** Lithium metal battery.

This is a primary cell, and thus is not of interest for manufacturers of automotive batteries. But it does produce a 3.4 V voltage. Also, it functions well because the electrolytes used are nonaqueous – a requirement when lithium is oxidized. They are either propylene carbonate or dimethoxyethane. As well, a small, nonstoichiometric amount of lithium perchlorate ($LiClO_4$) is in the electrolyte solution.

$LiCoO_2 + C_6 \rightarrow Li_{1-x}CoO_2 + Li_xC_6$ initial charging   **Figure 10.12:** Lithium ion
$Li_{1-x}CoO_2 + Li_xC_6 \rightarrow Li_{1-x+y}CoO_2 + Li_{x-y}C_6$ Discharge   battery.

The lithium ion battery is a secondary cell, and its charge and discharge cycle has been studied in detail. This is because it represents one of the best possibilities for a fully electric automobile. Note that the reactions shown in Figure 10.12 represent a form of lithium ion transport much more than a traditional redox couple of reactions. In the initial charging reaction, some lithium is lost from the lithium cobalt oxide and intercalated into the graphite. In the discharge portion of the cycle, a portion of the lithium that was intercalated is now lost, and again becomes art of the lithium cobalt oxide.

10.2.8 Other battery couples

The battery couples we have discussed here are very common, or have been, in modern society. Numerous other couples exist, as do numerous other battery types and sizes. It is fair to say that Table 10.1 is simply the tip of a very large iceberg. Interest remains keen in finding battery couples that do not involve any rare earth element (any lanthanide), and that utilize organic materials. The reason for this is the availability of rare earth elements, and their sourcing, which is discussed next [32].

10.3 Material sourcing and recycling for batteries

Numerous books and sources discuss battery chemistry, but some thought should also be given to where the materials originate that ultimately become batteries. Using the reaction chemistry that we have just seen in Section 10.2, we have found that the following elements and compounds are required [33]:

Lead, Pb
Lead refining is a mature industry, and is practiced in several countries. China currently has the largest production of the element, with Peru, Mexico, Australia, Russia, and the United States also having large annual outputs – in hundreds of thousands of metric tons.

Because lead refining has a considerable cost in terms of energy input, lead is often recycled. Lead-acid batteries have the lead plates recycled whenever possible.

Sulfuric acid, H_2SO_4
Sulfuric acid is made by the oxidation of elemental sulfur, or the oxidation of hydrogen sulfide (H_2S), and is the single largest commodity chemical production in the world. Thus, sources of sulfur are still considered plentiful, and production is well established.

Perhaps the larger problem with the life cycle of sulfuric acid is how to dispose of it in an environmentally friendly manner. Since the price of it hovers at approximately \$175 per ton, there is little incentive to recycle. Disappointingly, some sulfuric acid is still disposed of to its local environment.

Nickel, Ni
Nickel is found and refined in many places in the world. Since the uses of nickel are primarily for special, high-strength steels, recycling is focused on such large uses and end-use objects, and not on batteries.

Cadmium, Cd

Cadmium is routinely refined as a byproduct of zinc production. Thus, there is always an economic value to it. Cadmium has been recycled from NiCad batteries by at least one company in the United States.

Silver, Ag

Even though silver is a precious metal, or perhaps because it is, it is mined in many countries throughout the world. A precious metal such as gold is often measured in troy ounces because of its relative scarcity. But the USGS Mineral Commodity Summaries 2021 measures worldwide silver production in terms of metric tons because there is so much of it [33]. Leading producers are Mexico, Peru, and China. The United States does still produce some silver domestically. Beyond batteries, silver is used in luxury goods, bullion coins, as shown in Figure 10.13, and electronics.

Figure 10.13: Silver bullion coins.

Mercury, Hg

All mercury are ultimately sourced from HgS ores, called cinnabar. Because the use of mercury in batteries and in other applications is being phased out whenever possible, concerns about the availability of the element in the future are relatively small.

Zinc, Zn

Most zinc is refined for use in alloys, such as brass and bronze, as shown in Figure 10.14, and for galvanizing metals. Australia, China, and Peru remain the leading producers. Battery production remains a relatively small use for zinc. But because zinc is a part of the silver oxide battery, there is an economic incentive to recycle such batteries, that being the recovery of the silver.

Figure 10.14: Brass ingots.

Lithium, Li

Lithium is unevenly distributed around the Earth and is a relatively scarce element. Its scarcity is because there is really no nucleosynthetic pathway to it that is particularly favorable. Roughly 5% of the world's known lithium can be found in the region of Searles Lake, Nevada, in the United States. Most can be found in the Atacama High Desert in South America, and in mining operations in Australia. It is possible that a significant amount may also be found in Afghanistan. Since this nation has no history of large-scale mining, and because the long war there has made exploration dangerous, it is difficult to tell how extensive any deposits are.

Cobalt, Co

An enormous amount of the world's cobalt comes from the Congo, although several other countries are also net exporters. But cobalt is used much more in different alloys, especially in alloys that become aircraft parts, than is used in batteries. Recycling is thus a matter of recycling metal alloy parts.

Manganese, Mn

South Africa and Australia produce the majority of the world's manganese. Much more of the element is used each year in producing metal alloys, but its use in batteries remains an important one. Several of the elements listed here have some substitute, at least for some of their major uses. It is noteworthy that the USGS Mineral Commodity Summaries 2021 makes the terse statement about manganese: "Manganese has no satisfactory substitute in its major applications" [33]. Thus, recycling of manganese becomes important both currently and in the future.

References

[1] Advanced Lead Acid Battery Consortium. Website. (Accessed 28 July 2021, as: http://www.alabc.org/).

[2] Energy Storage Association. Website. (Accessed 28 July 2021, as: https://www.energystorage.org).

[3] Canadian Battery Association. Website. (Accessed 28 July 2021, as: https://canadianbatteryassociation.ca).

[4] The British and Irish Portable Battery Association. Website. (Accessed 28 July 2021, as: https://bipha.co.uk).

[5] Australian Battery Industry Association. Website. (Accessed 28 July 2021, as: https://www.abia.org.au).

[6] First National Battery – South Africa's Battery Manufacturer. Website. (Accessed 28 July 2021, as: https://www.battery.co.za).

[7] Eurobat. Website. (Accessed 28 July 2021, as: http://eurobat.org/).

[8] NAATBatt International. Website. (Accessed 28 July 2021, as: http://naatbatt.org).

[9] US Advanced Battery Consortium. Website. (Accessed 28 July 2021, as: http://www.uscar.org).

[10] PRBA: The Rechargeable Battery Association. Website. (Accessed 28 July 2021, as: http://www.prba.org/).

[11] European Battery Alliance. Website. (Accessed 28 July 2021, as: https://www.eba25.com).

[12] Recharge – Your European Battery Industry Association. Website. (Accessed 28 July 2021, as: http://www.rechargebatteries.org/).

[13] The Committee of Battery Technology. Website. (Accessed 28 July 2021, as: http://battery.electrochem.jp/index-e.html).

[14] Energy Storage Association. Website. (Accessed 28 July 2021, as: http://energystorage.org/).

[15] Battery Association of Japan. Website. (Accessed 28 July 2021, as: http://www.baj.or.jp/e/).

[16] Battery Council International. Website. (Accessed 28 July 2021, as: https://batterycouncil.org/).

[17] International Lithium Alliance. Website. (Accessed 28 July 2021, as: https://internationallithium.com/).

[18] Independent Battery Manufacturers Association. Website. (Accessed 28 July 2021, as: http://www.thebatteryman.com/).

[19] European Portable Battery Association. Website. (Accessed 28 July 2021, as: http://www.epbaeurope.net/).

[20] China Industrial Association of Power Sources. Website. (Accessed 28 July 2021, as: http://www.ciaps.org.cn/en/news/ShowInfo.aspx?NewsID=4).

[21] National Electrical Manufacturers Association. Website. (Accessed 28 July 2021, as: http://www.nema.org/pages/default.aspx).

[22] Battery Electrical Specialist Association. Website. (Accessed 28 July 2021, as: https://besabattery.com).

[23] National Electrical Equipment and Medical Imaging Manufacturers. Website. (Accessed 28 July 2021, as: http://www.nema.org/About/Pages/default.aspx).

[24] China Industrial Association of Power Sources. Website. (Accessed 28 July 2021, as: http://www.ciaps.org.cn/en/news/ShowInfo.aspx?NewsID=4).

[25] Battery, Recycling, and Manufacturing Associations. Website. (Accessed 28 July 2021, as: https://batterycouncil.org).

[26] Association of Battery Recyclers. Website. (Accessed 28 July 2021, as: http://www.associationofbatteryrecyclers.com/).

[27] Automotive Aftermarket Suppliers Association. Website. (Accessed 28 July 2021, as: https://www.aftermarketsuppliers.org/).

[28] International Lead Association. Website. (Accessed 28 July 2021, as: https://www.ila-lead.org/).

[29] Association of Battery Recyclers. Website. (Accessed 28 July 2021, as: https://www.associa tionofbatteryrecyclers.com).

[30] Automotive Aftermarket Suppliers Association. Website. (Accessed 28 July 2021, as: https:// www.aftermarketsuppliers.org).

[31] Call2Recycle. Website. (Accessed 28 July 2021, as: https://www.call2recycle.org).

[32] Electrochemical Energy Storage: Physics and Chemistry of batteries. R. Job., De Gruyter, 2020.

[33] U.S.G.S. Mineral Commodity Summaries 2021, downloadable as: http://pubs.usgs.gov/peri odicals/mcs2021/mcs2021.pdf.

Chapter 11
Energy harvesting

11.1 Energy harvesting – introduction

Every day an enormous amount of energy is created by the movement of people and animals, and by interactions of people with their immediate surroundings. This is usually in very small amounts or in very dispersed environments. Virtually all of that energy is lost to the local environment, and historically there have been no efforts to gather it. It may seem odd to consider finding ways to "collect" energy that is given off all around us – by people simply walking or by walking upstairs and downstairs or by riding stationary/exercise bicycles, for example – but that is the general idea and nature of energy harvesting. The broad idea of energy harvesting is that there are many places at which small amounts of energy are generated – and often wasted – and when collected, this can be put to some practical use. Current efforts have begun, aimed at collecting such energy in smaller devices which can store it, such as portable batteries.

Previous chapters have discussed the means of energy production that are large enough that regional, national, and international organizations exist to promote them and advocate for their use. Energy harvesting is not represented like this. Rather, while there are government agencies that do deal with it [1–3], there are many more universities and companies that have begun to dedicate some or all of their efforts at developing energy harvesting materials and devices that will become useful in a short period of time (months instead of years, e.g., or years instead of decades) [4–18]. It is likely that as some energy harvesting devices become products on the market, their presence will inspire the production of further related devices.

11.2 Potential areas for harvesting

It is difficult to make any complete and comprehensive list of sources and situations that can contribute to energy harvesting, simply because some of the newest will be left off. This is because the means of energy harvesting continue to change and evolve at a rapid pace. But the following are examples of energy harvesting that have already been proven to work and that could conceivably be expanded.

11.2.1 Traditional harvesting

Perhaps, the oldest, lowest tech type of energy harvesting is that of self-winding watches. A counterweight is placed inside the back watch casing, and the movement

https://doi.org/10.1515/9783110662276-011

of the wearer's arm is enough to make the weight swing, and rewind the watch while it is being worn. But this is hardly the type of scavenged energy that is thought of currently when we consider energy harvesting.

11.2.2 Bicycles

The idea of connecting a bicycle, or a stationary bicycle, to some turbine that then converts the energy a person puts into the bike while riding it to some amount of energy, such as electricity for a home, is another way to harvest personal energy. Since bicycles, including stationary bicycles, already have parts rotating about an axle, it is not hard to adapt them to turbine and generator to produce electricity [19].

This bicycle-to-energy idea has a certain appeal to people, and there have been attempts, some of them successful, to install such equipment in homes, sometimes some form of rack upon which a bicycle can be mounted and ridden. One can debate how useful this type of energy harvesting is, but the appeal is often to become more self-sustaining as a household, while at the same time engaging in physical exercise more on a daily basis. Using a stationary bicycle, as shown in Figure 11.1, or a bicycle in some support rack, in this way thus becomes a form of exercise and at the same time a means of harvesting energy.

Figure 11.1: Stationary bicycle.

11.2.3 Personal backpacks, walking options

A significant amount of energy is expended when a person walks for several kilometers, and possibilities exist for using the movement of walking or running as a source for energy.

The idea of wearing a backpack that can save and store the energy a person generates simply by walking and at the same time gently bouncing the backpack up and down may seem more like science fiction than science, yet such backpacks have already been designed and tested and are now available for purchase. In one design, motion caused by a person's movement generates an up-and-down movement in cylinders in the backpack frame. This movement of cylinders is then captured to store energy in some battery. There is interest in such backpacks by the United States Department of Defense, since soldiers already march significant distances. Once again, the motion of walking can be harnessed to charge batteries which can then be used with the equipment the soldier carries. In theory at least, this can be extended to anyone who carries such a backpack, such as hikers, outdoors enthusiasts, or commuters [20, 21].

11.2.4 Piezoelectric effect possibilities

The piezoelectric effect is the generation of a small amount of electricity when a crystalline material is deformed – which can mean electricity is generated when a material is stepped upon. Electric charge accumulates during the deformation – the compression or the stretch – shown schematically in Figure 11.2. A partial list of known piezoelectric materials include those shown in Table 11.1.

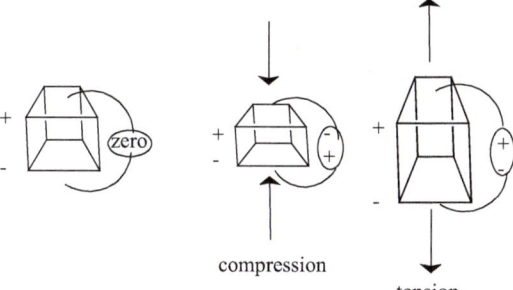

compression

tension

Figure 11.2: Piezoelectric effect.

Since the action of walking or running can compress a material, the possibility exists and has been examined in some detail that piezoelectric materials built into the sole or heel of a shoe, boot, or sock could generate electricity when a person walks.

Table 11.1: Common piezoelectric materials.

Name	Formula	Type
Aluminum nitride	AlN	Single crystal
Barium titanate	$BaTiO_3$	Single crystal/ceramic
Cadmium sulfide	CdS	Single crystal/ceramic
Gallium nitride	GaN	Ceramic
Lead titanate	$PbTiO_3$	Ceramic
Lead zirconate titanate	$Pb[Zr_xTi_{1-x}]O_3$	Ceramic
Polyvinylidene fluoride (PVDF)	$[CH_2CF_2]_n$	Polymeric
Rochelle's salt	$NaKC_4H_4O_6 \bullet 4H_2O$	Single crystal
Zinc oxide	ZnO	Ceramic

That can then be stored in batteries which are "plugged in" or removed from the upper part of any such foot wear.

In addition, piezoelectric materials can be integrated into a wide variety of small sensors, including those which are wearable [20, 21].

11.2.5 Stairs or rugs

People walk upstairs and downstairs constantly, especially in many public spaces, such as shopping malls and the entrances or exits of corporate and government buildings. The energy involved in placing a foot on a step can be harvested when special pads are placed on the steps, so that the up-down motion of stepping on the pads generates a small amount of energy per step.

Once again, this electricity gathered by movement on stairs or high traffic area rugs can be an example of some piezoelectric material being incorporated in the stair padding or floor covering [22].

11.2.6 Other possibilities

Many possibilities beyond those just mentioned exist for some form of energy harvesting, and several of these have been written about in the popular press [23, 24].

The idea has been proposed that harnessing the energy in revolving doors by connecting them to turbines and generators, then storing the produced energy in batteries, can be an effective way to harvest energy in malls, and by extension any location that sees high foot traffic and has a revolving door [25].

11.3 The future of energy harvesting

It sounds somewhat melodramatic to say that the future of energy harvesting is broad and promising, but this is indeed what is possible. For example:

1. For matting and carpeting to be produced on a large scale that can be laid down in high traffic areas of a home or business means that some portion of the power for that structure could be generated by people walking on the carpet [22].
2. For homes to be built with some connection to a stationary bicycle means that any home or apartment could have some of its power generated by residents exercising on such a bicycle [19].
3. For an entire generation of student backpacks to be equipped with pistons in their frames, so that walking motion can be harnessed to recharge batteries means less of a load on existing electrical grids, and ease of recharging batteries used for small electronics [22].
4. The use of revolving doors coupled to small turbines can mean energy savings for businesses where such doors are in use constantly [25].

All of these possibilities, as well as those which are under development now, represent ways to decrease the overall load on existing power grids. They also represent means by which people living off a grid can be more energy independent.

References

[1]   Energy Harvesting Consortium. Website. (Accessed 30 July 2021, as: http://nttdata-strategy. com/ehc/en/about).
[2]   U.S. Department of Energy. Office of Energy Efficiency & Renewable Energy. Website. (Accessed 30 July 2021, as: http://energy.gov/eere).
[3]   U.S. Department of Energy. Website. (Accessed 30 July 2021, as: http://energy.gov/eere/ amo/vibration-power-harvesting).
[4]   University of Alberta. Future Energy Systems. Website. (Accessed 30 July 2021, as: http:// futureenergysystems.ca/research/sustainability/wind/micro-scale-energy-harvesting-technology-in-remote-communities-of-alberta).
[5]   The American Ceramic Society. Website. (Accessed 30 July 2021, as: http://ceramics.org/ event/3rd-annual-energy-harvesting-society-meeting-archive).
[6]   The University of Melbourne. Energy Harvester. Website. (Accessed 30 July 2021, as: http:// chemical.eng.unimelb.edu.au/ellis/research/energy-harvesters).
[7]   Center for Energy Harvesting Materials and Systems. Virginia Polytechnic Institute. Website. (Accessed 30 July 2021, as: http://www.cehms.com/).
[8]   Power Sources Manufacturers Association. Website. (Accessed 30 July 2021, as: http://www. psma.com/).
[9]   Penn State at the Navy Yard. Website. (Accessed 30 July 2021, as: http://research.psu.edu/ energyharvesting).
[10] IDTechEx. Website. (Accessed 30 July 2021, as: https://www.idtechex.com/energy-harvesting -europe/show/en/).

[11] Energy Harvesting and Storage 2009. Website. (Accessed 30 July 2021, as: http://idtechex. com/energyharvestingandstorageeurope09/en/).

[12] Off Grid Energy Independence. Website. (Accessed 30 July 2021, as: www.offgridenergyind pendence.com).

[13] Fraunhofer Institute. Website. (Accessed 30 July 2021, as: https://www.fraunhofer.de/en. html).

[14] Micropelt. Website. (Accessed 30 July 2021, as: http://www.micropelt.com/).

[15] Perpetua. Website. (Accessed 30 July 2021, as: http://perpetuapower.com/).

[16] EnOcean. Website. (Accessed 30 July 2021, as: https://www.enocean.com/en/).

[17] Unlimited Energy. Website. (Accessed 30 July 2021, as: https://www.iter.org/).

[18] Mergeflow. Website. (Accessed 30 July 2021, http://mergeflow.com/emerging-technologies. html).

[19] J. Lange. Harvesting the Energy from Bicycles, 2017. Website. (Accessed 30 July 2021, as: http://large.stanford.edu/courses/2016/ph240/lange1/).

[20] M. Alhawari, B. Mohammad, H. Saleh. Energy Harvesting for Self-Powered Wearable Devices (Analog Circuits and Signal Processing), Springer, ISBN: 978-3-319-62577-5.

[21] Wearable Energy Storage Devices. A.M. Vinu Mohan, De Gruyter, 2022 (in press).

[22] T. Al-Qadhi, A. Al-Baser, F. Sammani, M. Elsayed, T. Jamil. Energy harvesting and power generation from stairs, AIP Conference Proceedings, 2035, 060008 (2018).

[23] The Power Plant of the Future Is Right in Your Home. Wired, Website. (Accessed 30 July 2021, as: http://wird.com/story/the-power-plant-of-the-future-is-right-in-your-home).

[24] This new device seems to pull electricity out of thin air: But don't expect it to power your city anytime soon. Popular Science, 20 February, 2020. Website. (Accessed 30 July 2021, as: http://popsci.com/story/science/air-gen-bacteria-moisture-electricity-renewable).

[25] P. Meshram, A. Kasurkar, P.V. Pullawar. Development in energy harvesting system using escalator & four way door mechanism. International Journal of Advanced Research in Science, Engineering and Technology. 2017, 4(4), 3708–3713.

Chapter 12
Energy and climate change

12.1 Introduction

The Earth appears to go through cycles, sometimes requiring millions of years, in which it warms and cools, often based on gases released to the atmosphere, and their ability to absorb or to block radiation from the Sun. One gas that has been used as something of a bellwether for warming or cooling of the planet, in terms of how much of it is present at any one time, is carbon dioxide (CO_2). Because of its linear shape, and because it contains three atoms, CO_2 is capable of vibrating in such a way that it can absorb energy, including energy from the Sun. Figure 12.1 shows the basic idea.

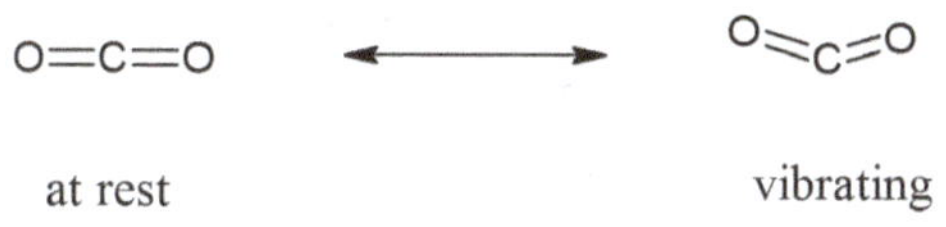

O=C=O $\longleftrightarrow$ O≈C≈O

at rest   vibrating

Figure 12.1: Carbon dioxide at rest and absorbing energy.

The vibration shows that a movement of the three atoms off the line of linearity for some measurable time can occur at very short time intervals and still absorb energy in the process. Repeated countless times on a global scale, this causes a significant increase of energy that is retained in any layer of the atmosphere where CO_2 is present. This cannot happen with either nitrogen gas, or oxygen gas, or argon, all of which are also components of the atmosphere. But two other gases that have bending modes, which means they can absorb energy in the atmosphere, are water vapor and methane. Figure 12.2 shows their Lewis structures.

Figure 12.2: Lewis structure of water and methane.

Methane is an especially potent greenhouse gas, because it has so many vibrational modes, and therefore has many ways to absorb solar energy. It is not difficult to compare the limited number of vibrations that CO_2 can undertake to the many that CH_4 can undergo. Viewed down a single C–H axis, as shown in Figure 12.3, there are at least three vibrations that can be made with any H–C–H bond, moving it off its 109.5° angle when at rest. All these are vibrations that can absorb energy.

Any discussion of climate change as a result of human activity and increased use of energy causes anything from discomfort to outrage, either on the part of those who think there is little that a single person can do to make some change for the good, or from those who insist that any current changes are part of a natural,

https://doi.org/10.1515/9783110662276-012

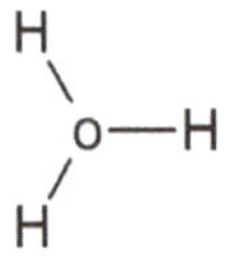

H
\
o—H
/
H

Figure 12.3: Methane viewed down a C–H bond.

planetary cycle. Yet all data from the past several decades indicate that the planet is warming. This data correspond well with the increased use of fossil fuels for power, as well as with the release of CO_2 into the atmosphere from industrial processes, such as iron production, as well as the release of any methane into the atmosphere. A large number of governmental and nongovernmental organizations exist to promote and advocate for some reduction of CO_2 and all greenhouse gases, as well as for the mitigation of climate change overall [1–57].

The use of coal and oil for energy has increased steadily for more than two centuries and has reached levels never before seen in history. Since the end product of the combustion of these two is CO_2 (among other gases such as carbon monoxide), this enhanced level of CO_2 means that more energy is being absorbed by the planet's atmosphere than at any other time in recorded history.

12.2 Types and sources of greenhouse gases

Numerous governmental and nongovernmental organizations have indicated that the presence of elevated levels of CO_2 in the atmosphere is a major source of material that is warming the planet. CO_2 levels have been monitored for several decades and have continued to rise with seasonal fluctuations. The levels of this one gas in the atmosphere have not been directly monitored since the start of the Industrial Revolution, but indirect methods have provided strong evidence for a direct correlation between the use of coal for energy generation and for iron refining, and the rise of CO_2 in the atmosphere.

There are certainly other greenhouse gases besides CO_2, some of which are much more potent in how they affect the atmosphere and climate. An examination includes the following gases.

12.2.1 Carbon dioxide (CO_2)

The means of vibration by which CO_2 traps energy and heat in the environment was just shown. Best estimates at the present are that levels of CO_2 are now at 415 ppm, up from 280 ppm in the year 1750 [13].

Table 12.1 shows a breakdown of greenhouse gases [11] and the human sector that produces them [11]. While this does not break the gases down to specific compounds, note that the listings for transportation and electricity are both based largely on the combustion of hydrocarbon-based fuels, and thus produce CO_2 as an end product.

Table 12.1: Greenhouse gas emissions.

Sector	Amount (%)	Examples
Transportation	29	Combustion of fossil fuels, cars, trucks, trains, planes, ships
Electricity	25	Coal and natural gas combustion, residential and commercial, all grid-connected
Industry	23	Metals production, chemical refining
Commercial and residential	13	HVAC, lighting
Agriculture	10	Fertilizer use, animal end-products

One sector that is actually an absorber of CO_2 is forestry. Managed forests can serve as what is called a "sink" for CO_2, meaning it is absorbed [11].

12.2.2 Methane (CH$_4$)

Although methane makes up a far smaller amount of gas in the atmosphere than CO_2, we have just seen that it is a potent greenhouse gas. At least one organization is active in determining means by which methane can be captured and used productively [12].

Methane is produced when animal wastes are concentrated, for instance, on large, industrial farms. But it is also a waste product from landfills. In the past, landfills were simply large holes in the ground. Now, however, landfills are made which are lined, which have piping systems to gather what is called leachate, and which can contain methane. Any methane which is captured can then be used, possibly as a heating fuel. Captured in this way, methane is no longer a fossil fuel.

12.2.3 Nitrous oxide (N$_2$O)

Nitrous oxide in the atmosphere has traditionally come from various soils, all of which are covered with vegetation. Its Lewis structure is shown in Figure 12.4. Like CO_2, it is linear and can absorb energy upon an induced bending at the central atom.

N≡N=O   Figure 12.4: Lewis structure of nitrous oxide.

The presence of over 1 billion internal combustion engines in automobiles, and their high compression of gases, means that more N_2O is now present in the atmosphere than has ever been present before. But it is also one of the products of the combustion of fossil fuels [11].

12.2.4 Ozone

The allotrope of oxygen that animal life cannot breathe, ozone (O_3), is a component of the greenhouse gas mix. Its Lewis structure is shown in Figure 12.5. Once again, it can absorb energy because the three-atom connection can vibrate.

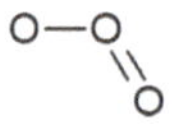

Figure 12.5: Ozone.

A naturally existing ozone layer covers the Earth – sometimes even called the ozonosphere – at approximately 15–25 km above the surface. When ozone is produced through man-made means, it becomes a problem, since it is at the Earth's surface and not in the upper atmosphere.

12.2.5 Chlorofluorocarbons

The presence of chlorofluorocarbons (CFCs) and of hydrochlorofluorocarbons is something relatively new in the atmosphere, as these are classes of compounds that simply do not exist in nature, and never have. Like the other gases we have seen, they are able to absorb energy from sunlight, and were engineered precisely because they do not readily break down into other materials. However, they were not designed to be part of Earth's atmosphere and have been part of the change that is occurring. Figure 12.6 shows two of the more common ones, although dozens have been created and tried as refrigerants.

Figure 12.6: Chlorofluorocarbons.

In the United States, the commercial production of CFCs was banned by the Toxic Substances Control Act. This was enacted in 1976, when the environmental damage caused by CFCs – their interaction and degradation of ozone – was first determined to be a problem that was expanding with expanding use of the gases.

Since this initial ban, almost 200 nations have banned the production and use of CFCs. This has resulted in what is believed to be a recovery of the natural ozone layer.

12.3 Governmental policies

Governmental bodies will always be involved in the debate about how to deal with climate change, simply because they have regulatory powers over the industries that exist within their borders, and because they are almost always in some way involved in the production, distribution, and regulation of energy. The problem of climate change and greenhouse gas emissions differs for different nations and is often based on the size of a nation. Using two extremes as examples, Singapore can enact regulation rather quickly and enforce it without having to have government employees travel huge distances, simply because the nation is quite small. On the other hand, Russia has a large mining and refining industry, both for fossil fuels and for metals. Enforcement of any regulations over such a large area is a much more difficult undertaking.

The problems of minimizing climate change and limiting greenhouse gas emissions can also be affected by the beliefs of those in office. Using the United States as an example, under the Obama administration, the Climate Action Plan was enacted to affect positive change in terms of greenhouse gas emissions, and thus pollution control [58]. Significant reversals of such policies were pursued under the Trump administration, starting from the first day of that administration.

An interface between government and industry occurs when the problems of climate change are intimately connected to the involvement of industry. An example is when significant amounts of materials such as steel are required for a nation's economic health. The production of it for use in automobiles, railroads, shipping, and defense means it is not something that can be curtailed immediately. Economically, such would throw large numbers of people out of work. Chemically, there is no better way to produce steel than is currently in place – which produces large amounts of CO_2. Additionally, such an industry may have significant lobbying power to its government. This combines to make it difficult to decrease the amount of pollutants generated by such an industry.

While it may seem short-sighted of politicians to deny evidence of climate change, it should be remembered that many governmental leaders are elected by the people they govern, and thus think in the shorter term, often no farther than their next election cycle. Perhaps obvious examples are those of a politician elected to represent an area which employs numerous coal miners or mill workers. Those miners and workers must eat today and feed their families today; and thus it is very difficult to simply tell them to get other jobs, or to get other energy-related jobs and careers.

12.3.1 Tax incentives

One means by which governments attempt to control the emission of carbon from industrial concerns is by some form of tax reduction or incentive. As just mentioned, it is understood that an industrial concern cannot simply stop producing whatever it makes in order to cease producing greenhouse gases, principally CO_2. Thus, the offer is made that taxes can be lowered for a company or site that reduces its carbon emissions.

12.3.2 Regulations

Numerous governments have made regulations regarding the production of greenhouse gases, in the hopes of limiting how much they are produced. The just-mentioned Climate Action Plan is one example [58]. A much earlier one is the United States Clean Air Act [59], although this was preceded by the United Kingdom's Clean Air Act of 1956 [60]. First enacted in 1963, the US act has been amended in 1965, 1967, 1970, 1977, and 1990.

Both acts, as well as others like them, were enacted because in the 1950s, the air quality in some US cities, as well as in London, was so bad it was deemed toxic. What is sometimes called "The Smog of 1952" in London made walking outside in public dangerous, simply because breathing the air for any length of time caused problems. In the United States, some cities were so polluted that drivers had to keep their lights on at any time, day or night. Pittsburgh, which lies in a valley at the junction of three rivers, and which was a city with numerous steel mills, is one example.

These attempts to improve the air quality, and thus people's quality of life, have largely been successful over the course of time.

12.4 Public perceptions and opinion

The perception by the general public as to whether or not climate change is occurring is often not decided by the presentation of scientific facts, but by how the information is presented to the public by the news media, or by political leaders. When political leaders deny any change, such public pronouncements tend to reinforce the idea among many that since a figure in authority has denied it, and since this cannot be seen or felt in their immediate locale, it must not be occurring, or must not be of concern to them.

The simple fact that climate change is not a visible phenomenon means that many people will choose to disbelieve that it is occurring, or will believe that any changes are part of the natural cycles of the Earth's weather. Unlike the problem of excess solid or liquid waste in landfills or bodies of water, which is visible, unsightly, and at times smelly, climate change is only quantifiable through somewhat indirect, invisible means.

The following are considered signs of climate change by many, but as natural phenomena by others:

1. Extremes of heat and of cold
2. Change in sea level
3. Melting of glaciers
4. Wildfires
5. Deforestation
6. Loss of coral reef life

In each of the six examples above, those who believe that climate change is occurring point to such phenomena as proof of it. Deniers tend to believe that all of these are part of Earth's natural cycles and that we are simply seeing one swing in a much greater series of cyclic swings. However, both sides of this large debate do tend to be in agreement on the very simple fact that the Earth remains the only habitable planet we have and that conserving its resources for future use is at least prudent, and at best a wise investment in the future.

References

[1]   350.org. Website. (Accessed 30 July 2021, as: https://350.org/).
[2]   Sierra Club Foundation. Website. (Accessed 20 May 2019, as: https://act.sierraclub.org/).
[3]   Environmental Defense Fund. Website. (Accessed 20 May 2019, as: https://www.edf.org/).
[4]   Union of Concerned Scientists. Website. (Accessed 30 July 2021, as: https://secure.ucsusa.org/).
[5]   Greenpeace. Website. (Accessed 20 May 2019, as: https://www.greenpeace.org/usa/).
[6]   Earth Justice. Website. (Accessed 20 May 2019, as: https://earthjustice.org/).
[7]   Center for Climate and Energy Solutions. Website. (Accessed 30 July 2021, as: https://www.c2es.org/).
[8]   The Solutions Project. Website. (Accessed 20 May 2019, as: https://thesolutionsproject.org/).
[9]   Nature Conservancy. Website. (Accessed 20 May 2019, as: https://www.nature.org/en-us/).
[10] National Resources Defense Council. Website. (Accessed 20 May 2019, as: https://act.nrdc.org/).
[11] U.S. Environmental Protection Agency. Website. (Accessed 31 July 2021, as: https://www.epa.gov).
[12] Global Methane Initiative. Website. (Accessed 31 July 2021, as: https://globalmethane.org).
[13] NOAA Education. Website. (Accessed 31 July 2021, as: https://noaa.gov/education).
[14] Intergovernmental Panel on Climate Change (IPCC). Website. (Accessed 31 July 2021, as: https://www.ipcc.ch).
[15] National Center for Atmospheric Research (NCAR). Website. (Accessed 31 July 2021, as: https://ncar.ucar.edu).

[16]   Center for Remote Sensing of Ice Sheets (CReSIS). Website. (Accessed. 31 July 2021, as: https://cresis.ku.edu).

[17]   National Climate Data Center (NCDC). Website. (Accessed 31 July 2021, as: https://www.ncdc.noaa.gov).

[18]   World Meteorological Organization. Website. (Accessed 31 July 2021, as: https://public.wmo.int).

[19]   United Nations Environmental Programme (UNEP), Climate Change. Website. (Accessed 31 July 2021, as: https://www.unep.org).

[20]   Food and Agriculture Organization (FAO) of the United Nations – Climate Change. Website. (Accessed 31 July 2021, as: https://fao.org/climate-change/en).

[21]   National Snow and Ice Data Center. Website. (Accessed 31 July 2021, as: https://nsidc.org).

[22]   International Geosphere-Biosphere Programme (IGBP). Website. (Accessed 31 July 2021, as: http://www.igbp.net).

[23]   Canadian Council on Ecological Areas. Website. (Accessed 31 July 2021, as: https://ccea-ccae.org).

[24]   Land Trust Alliance of British Columbia. Website. (Accessed 31 July 2021, as: https://www.landtrustalliance.org).

[25]   Canadian Parks & Wilderness Society (CPAWS). Website. (Accessed 31 July 2021, as: https://cpaws.org).

[26]   Canadian Wildlife Federation. Website. (Accessed 31 July 2021, as: https://www.cwf-fcf.org).

[27]   Conservation Ontario. Website. (Accessed 31 July 2021, as: https://conservationontario.ca).

[28]   Ducks Unlimited Canada. Website. (Accessed 31 July 2021, as: https://www.ducks.ca).

[29]   British Ecological Society. Website. (Accessed 31 July 2021, as: https://www.britishecologicalsociety.org).

[30]   National Association for Environmental Education. Website. (Accessed 31 July 2021, as: https://naee.org.uk).

[31]   Energy Saving Trust. Website. (Accessed 31 July 2021, as: https://energysavingtrust.org.uk).

[32]   Friends of the Earth. Website. (Accessed 31 July 2021, as: https://friendsoftheearth.uk).

[33]   Ethical Consumer. Website. (Accessed 31 July 2021, as: https://www.ethicalconsumer.org).

[34]   The Wildlife Trusts. Website. (Accessed 31 July 2021, as: https://www.wildlifetrusts.org).

[35]   European Environmental Bureau. Website. (Accessed 31 July 2021, as: https://eeb.org).

[36]   Seas at Risk. Website. (Accessed 31 July 2021, as: https://seas-at-risk.org).

[37]   Coalition Clean Baltic. Website. (Accessed 31 July 2021, as: https://ccb.se).

[38]   World Wide Fund, European Policy Office. Website. (Accessed 31 July 2021, as: https://www.wwf.eu).

[39]   Naturefriends International. Website. (Accessed 31 July 2021, as: https://www.nf-int.org).

[40]   Climate Action Network Europe. Website. (Accessed 31 July 2021, as: https://caneurope.org).

[41]   Taiga Rescue Network. Website. (Accessed 31 July 2021, as: https://web.archive.org/web/20130406214022/http://taigarescue.org).

[42]   GEN Europe. Website. (Accessed 31 July 2021, as: https://gen-europe.org)

[43]   European Forum on Nature Conservation and Pastoralism. Website. (Accessed 31 July 2021, as: http://www.efncp.org).

[44]   EUCC – the Coastal Union. Website. (Accessed 31 July 2021, as: https://www.eucc.net).

[45]   Climate Institute of Australia. Website. (Accessed 31 July 2021, as: https://australianinstitute.org.au).

[46]   The Climate Group. Website. (Accessed 31 July 2021, as: https://www.theclimategroup.org).

[47]   Climate Emergency Action Alliance. Website. (Accessed 31 July 2021, as: https://www.climatealliance.org).

[48]   Climate Council. Website. (Accessed 31 July 2021, as: https://www.climatecouncil.org.au).

[49] Clean Energy Regulator. Website. (Accessed 31 July 2021, as: http://www.cleanenergyregula tor.gov.au).

[50] Tauranga Carbon Reduction Group. Website. (Accessed 31 July 2021, as: http://www.nzcan.org).

[51] Lawyers for Climate Action New Zealand. Website. (Accessed 31 July 2021, as: https://www. lawyersforclimateaction.nz).

[52] Wildlife and Environment Society of South Africa (WESSA). Website. (Accessed 31 July 2021, as: https://wessa.org.za).

[53] Wilderness Foundation Africa (WFA). Website. (Accessed 31 July 2021, as: https://www.wilder nessfoundation.co.za).

[54] Botanical Society of South Africa. Website. (Accessed 31 July 2021, as: https://botanicalsoci ety.org.za).

[55] World Wide Fund for Nature (WWF) South Africa. Website. (Accessed 31 July 2021, as: https:// www.wwf.org.za).

[56] United Nations. UNFCCC. "What is the United Nations Framework Convention on Climate Change?" Website. (Accessed 31 July 2021, as: https://unfccc.int).

[57] Kyoto Protocol. Website. (Accessed 31 July 2021, as: https://unfccc.int/resource/docs/ conkp/kpeng.pdf).

[58] Climate Action Plan. Website. (Accessed 31 July 2021, as: https://obamawhitehouse.archives. gov/the-record/climate).

[59] Summary of the Clean Air Act. Website. (Accessed 31 July 2021 as: https://epa.gov/laws-regulations/summary-clean-air-act).

[60] Clean Air Act of 1956. Website. (Accessed 31 July 2021 as: https://legislation.gov.uk/ukpga/ Eliz2/4-5/52/enacted).

Index

https://doi.org/10.1515/9783110662276-013